AF689481

ENTOMOLOGY: AN INTRODUCTION

Edited by
Nada Ben Abdallah

ArclerPress

Entomology: An Introduction

Edited by Nada Ben Abdallah

© 2017

Edition 2018

ArclerPress

Published by
Arcler Press LLC
708 3rd Avenue, 6th Floor
New York
NY 10017
United States of America
www.arclerpress.com

ISBN: 978-1-68094-537-9 (Hardback)

Printed in United States of America

Library of Congress Control Number: 2017933109

This book contains information obtained from highly regarded resources. Some content has been published with permission under Creative Commons License or similar. A Wide variety of references are listed. Reasonable efforts have been made to publish reliable data and views articulated in the chapters are those of the individual contributors, and not necessarily those of the editors or publishers. Editors or publishers are not responsible for the accuracy of the information in the published chapters or consequences of their use. The publisher believes no responsibility for any damage or grievance to the persons or property arising out of the use of any materials, instructions, methods or thoughts in the book. The editors and the publisher have attempted to trace the copyright holders of all material reproduced in this publication and apologize to copyright holders if permission has not been obtained. If any copyright holder has not been acknowledged, please write to us so we may rectify.

Notice: Registered trademark of products or corporate names are used only for explanation and identification without intent of infringement. Although care has been taken to check accuracy of formulas and procedures, the detailed methods should be tested further on a small scale before being adopted commercially.

Arcler Press publishes wide variety of books and eBooks. For more information about Arcler Press and its products, visit our website at www.arclerpress.com.

About the Editor

Nada Ben Abdallah

Nada graduated from the University of Zurich in 2010 and holds a PhD in neuroscience. Her research focuses on elucidating the processes of learning and memory as well as the affective processes and their underlying neural mechanisms. Her work so far involved normal and diseased animal models with a main interest in neurodegenerative diseases. She is currently working as a freelance researcher and science editor.

Table of Contents

2 ARACHNOLOGY

3 ETHNOENTOMOLOGY

4 FORENSIC ENTOMOLOGY

5 MEDICAL ENTOMOLOGY

6 ECONOMIC ENTOMOLOGY

List of Abbreviations

AFRAS	African Arachnological Society
AAS	American Arachnological Society
ATS	American Tarantula Society
ARABEL	Arachnologia Belgica - Belgian Arachnological Society
ASJ	Arachnological Society of Japan
ASA	Asian Society of Arachnology
BAS	British Arachnological Society
BTS	British Tarantula Society
ESA	European Society of Arachnology
ISA	International Society of Arachnology
IAS	Iranian Arachnological Society
PMI	Post mortem interval

Preface

Entomology is the science that deals with everything about insects. The field of entomology often interacts with a wide range of scientific activities, including ecology, biochemistry, anthropology, developmental biology. Insects are involved in virtually every part of our lives; they are pests that eat our food, our houses, our animals, and are vectors that spread sickness and disease. But insects are not all bad! Many insects are beneficial pollinators, decomposers of dead materials, and useful in the biocontrol of unwanted pests. Entomologists study insects to help us manage pests, or learn how to better use them to our advantage. Entomologists are scientists who study insects and how they interact with other species, humans, and the environment. The scientific classification of an insect is an animal that has segmented body parts. There are over one million identified species with most entomologists estimating there to be an additional six to eight more unidentified species. Insect ecology, studies only insects and their relationships with their environment, while insect morphology, studies insect body parts and functions. Insect pathology focuses on how insect sicknesses become disease agents. Insect toxicology analyzes how chemicals and insecticides affect insect physiological functions in order to improve pest management and agricultural production. Medical entomologists study how to prevent insects from harming humans, thus focusing on fatal diseases such as dengue and malaria. The subject does not exist in a bubble where it is useful only to itself; it is part of a wider environmental study of our natural ecology and human geography - insects are affected by the climate and results of studies can tell us much about what is going on in the world around us. We discover new uses for studies in entomology all the time and there are already several established disciplines where insects and insect remains are a powerful type of evidence.

Content Coverage

Chapter One aims to insects physiology and morphology; a class of invertebrates within the arthropod phylum that have a chitinous exoskeleton, a three-part body (head, thorax and abdomen), three pairs of jointed legs, compound eyes, and one pair of antennae.

Chapter Two presents the study of spiders and related animals such as scorpions, pseudo scorpions, and harvestmen, collectively called arachnids.

Chapter Three focuses primarily on the roles of insects in culture that span different aspects of human life, whether analyzed academically or more generally.

Chapter Four is about forensic entomology that deals with the use of the insects, and their arthropod relatives that inhabit decomposing remains to aid legal investigations. The broad field of forensic entomology is commonly broken down into three general areas: medicolegal, urban, and stored product pests.

Chapter Five deals with insects that cause disease or that serve as vectors of organisms that cause disease in humans.

Chapter Six emphasizes on the economic entomology, which involves the study of insects that benefit or harm humans, domestic animals, and crops. Insects that cause losses are termed as pests. Some species can cause indirect damage by spreading diseases and these are termed as vectors.

Chapter Seven presents the study of myriapods which includes centipedes and millipedes. The field of myriapodology can also cover other myriapods such as pauropods and symphylans. Those who study myriapods are myriapodologists.

Chapter Eight offers the study of butterflies and moths (the lepidoptera),where butterfly is a flying insect of the order lepidoptera , distinguished from moths by their diurnal activity and generally brighter colouring.

CHAPTER 1

Insects: Evolution and Physiology

Chapter Contents

Insects are a large group of arthropods (segmented animals) that make up to more than half of animal biodiversity on Earth. Over a million species have been described, with recent estimates of the total number of species reaching up to 10 million. Only about a million non-insect animal species have been described. Common insect species include the dragonfly, praying mantis, grasshopper, butterfly, moth, fly, true bugs, beetles, bees, wasps, and ants. Insects have been around since at least the Devonian Period, about 400 million years ago, when the first vascular plants appeared on land. Insects are primary terrestrial – their close relatives, crustaceans, are better adapted to the seas.

Insects have three main body sections; a head, thorax, and abdomen. The head features compound eyes, sensory antennae, and modified appendages, which serve as mouthparts. The abdomen contains most of the organs. Insects lack both lungs and a closed circulatory system. Oxygen exchange is accomplished through trachea and direct diffusion with the outside. Insect "blood" is called hemolyph. Hemolyph distributes oxygen to organs simply by washing over them, rather than using blood vessels, as in mammals.

INTRODUCTION

Insects, are a class of invertebrates within the arthropod phylum that have a chitinous exoskeleton, a three-part body (head, thorax and abdomen), three pairs of jointed legs, compound eyes and one pair of antennae. They are the most diverse group of animals on the planet, including more than a million described species and representing more than half of all known living organisms. The number of extant species is estimated at between six and ten million, and potentially represent over 90% of the differing animal life forms on Earth. Insects may be found in nearly all environments, although only a small number of species reside in the oceans, a habitat dominated by another arthropod group, crustaceans.

The life cycles of insects vary but most hatch from eggs. Insect growth is constrained by the inelastic exoskeleton and development involves a series of molts. The immature stages can differ from the adults in structure, habit and habitat, and can include a passive pupal stage in those groups that undergo 4-stage metamorphosis (see holometabolism). Insects that undergo 3-stage metamorphosis lack a pupal stage and adults develop through a series of nymphal stages. The higher level relationship of the Hexapoda is unclear. Fossilized insects of enormous size have been found from the Paleozoic Era, including giant dragonflies with wingspans of 55 to 70 cm (22–28 in). The most diverse insect groups appear to have coevolved with flowering plants.

Adult insects typically move about by walking, flying or sometimes swimming *(see* ***‹See Tfd'*** *§ Locomotion below)*. As it allows for rapid yet stable movement, many insects adopt a tripedal gait in which they walk with their legs touching the ground in alternating triangles. Insects are the only invertebrates to have evolved flight. Many insects spend at least part of their lives under water, with larval adaptations that include gills, and some adult insects are aquatic and have adaptations for swimming. Some species, such as water striders, are capable of walking on the surface of water. Insects are mostly solitary, but some, such as certain bees, ants and termites, are social and live in large, well-organized colonies. Some insects, such as earwigs, show maternal care, guarding their eggs and young. Insects can communicate with each other in a variety of ways. Male moths can sense the pheromones of female moths over great distances. Other species communicate with sounds: crickets stridulate, or rub their wings together, to attract a mate and repel other males. Lampyridae in the beetle order communicate with light.

Humans regard certain insects as pests, and attempt to control them using insecticides and a host of other techniques. Some insects damage crops by feeding on sap, leaves or fruits. A few parasitic species are pathogenic. Some insects perform complex ecological roles; blow-flies, for example, help consume carrion but also spread diseases. Insect pollinators are essential to the life-cycle of many flowering plant species on which most organisms, including humans, are at least partly dependent; without them, the terrestrial portion of the biosphere (including humans) would be devastated. Many other insects are considered ecologically beneficial as predators and a few provide direct economic benefit. Silkworms and bees have been used extensively by humans for the production of silk and honey, respectively. In some cultures, people eat the larvae or adults of certain insects.

ETYMOLOGY

The word "insect" comes from the Latin word *insectum*, meaning "with a notched or divided body", or literally "cut into", from the neuter singular perfect passive participle of *insectare*, "to cut into, to cut up", from *in-* "into" and *secare* "to cut"; because insects appear "cut into" three sections. Pliny the Elder introduced the Latin designation as a loan-translation of the Greek word ντομος (*éntomos*) or "insect" (as in entomology), which was Aristotle's term for this class of life, also in reference to their "notched" bodies. "Insect" first appears documented in English in 1601 in Holland's translation of Pliny. Translations of Aristotle's term also form the usual word for "insect" in Welsh (trychfil, from *trychu* "to cut" and *mil*, "animal"), Serbo-Croatian (*zareznik*, from *rezati*, "to cut"), Russian (насекомое *nasekomoje*, from *seč'/-sekat'*, "to cut"), etc.

PHYLOGENY AND EVOLUTION

Evolution has produced enormous variety in insects. Pictured are some of the possible shapes of antennae.

The evolutionary relationship of insects to other animal groups remains unclear.

Although traditionally grouped with millipedes and centipedes—possibly on the basis of convergent adaptations to terrestrialisation—evidence has emerged favoring closer evolutionary ties with crustaceans. In the Pancrustacea theory, insects, together with Entognatha, Remipedia, and Cephalocarida, make up a natural clade labeled Miracrustacea.

A report in November 2014 unambiguously places the insects in one clade, with the crustaceans and myriapods, as the nearest sister clades. This study resolved insect phylogeny of all extant insect orders, and provides "a robust phylogenetic backbone tree and reliable time estimates of insect evolution."

Other terrestrial arthropods, such as centipedes, millipedes, scorpions, and spiders, are sometimes confused with insects since their body plans can appear similar, sharing (as do all arthropods) a jointed exoskeleton. However, upon closer examination, their features differ significantly; most noticeably, they do not have the six-legged characteristic of adult insects.

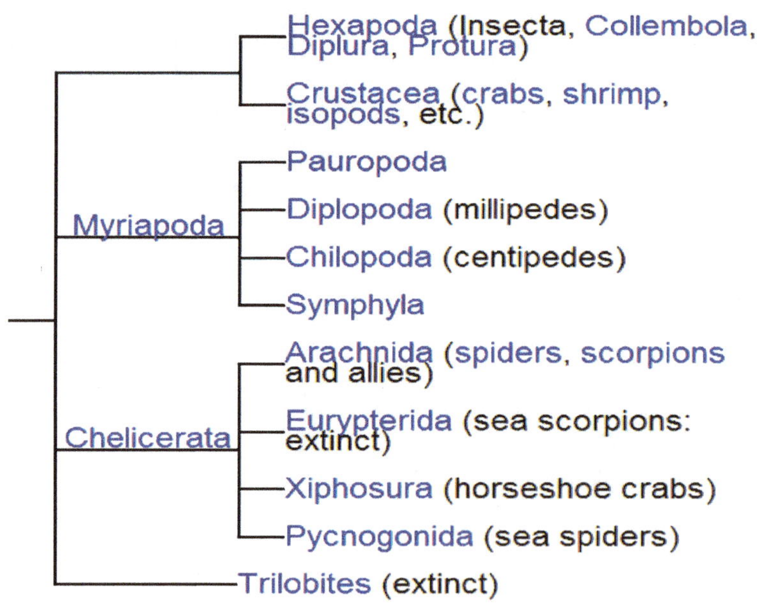

Cladogram of living insect groups, with numbers of species in each group. The Apterygota, Palaeoptera, and Exopterygota are possibly paraphyletic groups.

The higher-level phylogeny of the arthropods continues to be a matter of debate and research. In 2008, researchers at Tufts University uncovered what they believe is the world's oldest known full-body impression of a primitive flying insect, a 300 million-year-old specimen from the Carboniferous period. The oldest definitive insect fossil is the Devonian *Rhyniognatha hirsti*, from the 396-million-year-old Rhynie chert. It may have superficially resembled a modern-day silverfish insect. This species already possessed dicondylic mandibles (two articulations in the mandible), a feature associated with winged insects, suggesting that wings may already have evolved at this time. Thus, the first insects probably appeared earlier, in the Silurian period.

Four super radiations of insects have occurred: beetles (evolved about 300 million years ago), flies (evolved about 250 million years ago), and moths and wasps (evolved about 150 million years ago). These four groups account for the majority of described species. The flies and moths along with the fleas evolved from the Mecoptera.

The origins of insect flight remain obscure, since the earliest winged insects currently known appear to have been capable fliers. Some extinct insects had an additional pair of winglets attaching to the first segment of the thorax, for a total of three pairs. As of 2009, no evidence suggests the insects were a particularly successful group of animals before they evolved to have wings.

Late Carboniferous and Early Permian insect orders include both extant groups, their stem groups, and a number of Paleozoic groups, now extinct. During this era, some giant dragonfly-like forms reached wingspans of 55 to 70 cm (22 to 28 in), making them far larger than any living insect. This gigantism may have been due to higher atmospheric oxygen levels that allowed increased respiratory efficiency relative to today. The lack of flying vertebrates could have been another factor. Most extinct orders of insects developed during the Permian period that began around 270 million years ago. Many of the early groups became extinct during the Permian-Triassic extinction event, the largest mass extinction in the history of the Earth, around 252 million years ago.

The remarkably successful Hymenoptera appeared as long as 146 million years ago in the Cretaceous period, but achieved their wide diversity more recently in the Cenozoic era, which began 66 million years ago. A number of highly successful insect groups evolved in conjunction with flowering plants, a powerful illustration of coevolution.

Many modern insect genera developed during the Cenozoic. Insects from this period on are often found preserved in amber, often in perfect condition. The body plan, or morphology, of such specimens is thus easily compared with modern species. The study of fossilized insects is called paleoentomology.

Evolutionary Relationships

Insects are prey for a variety of organisms, including terrestrial vertebrates. The earliest vertebrates on land existed 400 million years ago and were large amphibious piscivores. Through gradual evolutionary change, insectivory was the next diet type to evolve.

Insects were among the earliest terrestrial herbivores and acted as major selection agents on plants. Plants evolved chemical defenses against this herbivory and the insects, in turn, evolved mechanisms to deal with plant toxins. Many insects make use of these toxins to protect themselves from their predators. Such insects often advertise their toxicity using warning colors. This successful evolutionary pattern has also been used by mimics. Over time, this has led to complex groups of coevolved species. Conversely, some interactions between plants and insects, like pollination, are beneficial to both organisms. Coevolution has led to the development of very specific mutualisms in such systems.

Taxonomy

Traditional morphology-based or appearance-based systematics have usually given the Hexapoda the rank of superclass, and identified four groups within it: insects (Ectognatha), springtails (Collembola), Protura, and Diplura, the latter three being grouped together as the Entognatha on the basis of internalized mouth parts. Supraordinal relationships have undergone numerous changes with the advent of methods based on evolutionary history and genetic data. A recent theory is that the Hexapoda are polyphyletic (where the last common ancestor was not a member of the group), with the entognath classes having separate evolutionary histories from the Insecta. Many of the traditional appearance-based taxa have been shown to be paraphyletic, so rather than using ranks like subclass, superorder, and infraorder, it has proved better to use monophyletic groupings (in which the last common ancestor is a member of the group). The following represents the best-supported monophyletic groupings for the Insecta.

Insects can be divided into two groups historically treated as subclasses: wingless insects, known as Apterygota, and winged insects, known as Pterygota. The Apterygota consist of the primitively wingless order of the silverfish (Thysanura). Archaeognatha make up the Monocondylia based on the shape of their mandibles, while Thysanura and Pterygota are grouped together as Dicondylia. The Thysanura themselves possibly are not monophyletic, with the family Lepidotrichidae being a sister group to the Dicondylia (Pterygota and the remaining Thysanura).

Paleoptera and Neoptera are the winged orders of insects differentiated by the presence of hardened body parts called sclerites, and in the Neoptera, muscles that allow their wings to fold flatly over the abdomen. Neoptera can further be divided into incomplete metamorphosis-based (Polyneoptera and Paraneoptera) and complete metamorphosis-based groups. It has proved difficult to clarify the relationships between the orders in Polyneoptera because of constant new findings calling for revision of the taxa. For example, the Paraneoptera have turned out to be more closely related to the Endopterygota than to the rest of the Exopterygota. The recent molecular finding that the traditional louse orders Mallophaga and Anoplura are derived from within Psocoptera has led to the new taxon Psocodea. Phasmatodea and Embiidina have been suggested to form the Eukinolabia. Mantodea, Blattodea, and Isoptera are thought to form a monophyletic group termed Dictyoptera.

The Exopterygota likely are paraphyletic in regard to the Endopterygota. Matters that have incurred controversy include Strepsiptera and Diptera grouped together as Halteria based on a reduction of one of the wing pairs – a position not well-supported in the entomological community. The Neuropterida are often lumped or split on the whims of the taxonomist. Fleas are now thought to be closely related to boreid mecopterans. Many questions remain in the basal relationships amongst endopterygote orders, particularly the Hymenoptera.

The study of the classification or taxonomy of any insect is called systematic entomology. If one works with a more specific order or even a family, the term may also be made specific to that order or family, for example systematic dipterology.

DIVERSITY

Though the true dimensions of species diversity remain uncertain, estimates range from 2.6–7.8 million species with a mean of 5.5 million. This probably represents less than 20% of all species on Earth, and with only about 20,000 new species of all organisms being described each year, most species likely will remain undescribed for many years unless species descriptions increase in rate. About 850,000–1,000,000 of all described species are insects. Of the 24 orders of insects, four dominate in terms of numbers of described species, with at least 3 million species included in Coleoptera, Diptera, Hymenoptera and Lepidoptera. A recent study estimated the number of beetles at 0.9–2.1 million with a mean of 1.5 million.

Comparison of the estimated number of species in the four most speciose insect orders

	Described species	Average description rate (species per year)	Publication effort
Coleoptera	300,000–400,000	2308	0.01
Lepidoptera	110,000–120,000	642	0.03
Diptera	90,000–150,000	1048	0.04
Hymenoptera	100,000–125,000	1196	0.02

MORPHOLOGY AND PHYSIOLOGY

External

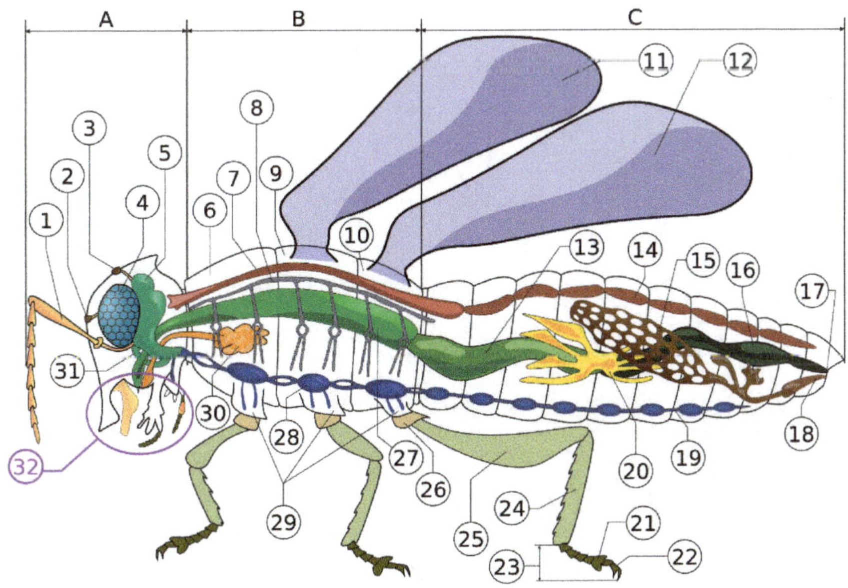

Insect morphology

A- Head **B**- Thorax **C**- Abdomen

1. antenna
2. ocelli (lower)
3. ocelli (upper)
4. compound eye
5. brain (cerebral ganglia)
6. prothorax
7. dorsal blood vessel
8. tracheal tubes (trunk with spiracle)
9. mesothorax
10. metathorax
11. forewing
12. hindwing
13. mid-gut (stomach)
14. dorsal tube (Heart)
15. Ovary
16. hind-gut (intestine, rectum & anus)
17. Anus
18. Oviduct
19. nerve chord (abdominal ganglia)
20. Malpighian tubes
21. tarsal pads
22. Claws
23. tarsus
24. tibia
25. femur
26. trochanter
27. fore-gut (crop, gizzard)
28. thoracic ganglion
29. coxa
30. salivary gland
31. subesophageal ganglion
32. mouthparts

Insects have segmented bodies supported by exoskeletons, the hard outer covering made mostly of chitin. The segments of the body are organized into three distinctive but interconnected units, or tagmata: a head, a thorax and an abdomen. The head supports a pair of sensory antennae, a pair of compound eyes, and, if present, one to three simple eyes (or ocelli) and three sets of variously modified appendages that form the mouthparts. The thorax has six segmented legs—one pair each for the prothorax, mesothorax and the metathorax segments making up the thorax—and, none, two or four wings. The abdomen consists of eleven segments, though in a few species of insects, these segments may be fused together or reduced in size. The abdomen also contains most of the digestive, respiratory, excretory and reproductive internal structures. Considerable variation and many adaptations in the body parts of insects occur, especially wings, legs, antenna and mouthparts.

Segmentation

The head is enclosed in a hard, heavily sclerotized, unsegmented, exoskeletal head capsule, or epicranium, which contains most of the sensing organs, including the antennae, ocellus or eyes, and the mouthparts. Of all the insect orders, Orthoptera displays the most features found in other insects, including the sutures and sclerites.Here, the vertex, or the apex (dorsal region), is situated between the compound eyes for insects with a hypognathous and opisthognathous head. In prognathous insects, the vertex is not found between the compound eyes, but rather, where the ocelli are normally. This is because the primary axis of the head is rotated 90° to become parallel to the primary axis of the body. In some species, this region is modified and assumes a different name.

The thorax is a tagma composed of three sections, the prothorax, mesothorax and the metathorax. The anterior segment, closest to the head, is the prothorax, with the major features being the first pair of legs and the pronotum. The middle segment is the mesothorax, with the major features being the second pair of legs and the anterior wings. The third and most posterior segment, abutting the abdomen, is the metathorax, which features the third pair of legs and the posterior wings. Each segment is dilineated by an intersegmental suture. Each segment has four basic regions. The dorsal surface is called the tergum (or *notum*) to distinguish it from the abdominal terga. The two lateral regions are called the pleura (singular: pleuron) and the ventral aspect is called the sternum. In turn, the notum of the prothorax is called the pronotum, the notum for the mesothorax is called the mesonotum and the notum for the metathorax is called the metanotum. Continuing with this logic, the mesopleura and metapleura, as well as the mesosternum and metasternum, are used.

The abdomen is the largest tagma of the insect, which typically consists of 11–12 segments and is less strongly sclerotized than the head or thorax. Each segment of the abdomen is represented by a sclerotized tergum and sternum. Terga are separated from each other and from the adjacent sterna or pleura by membranes. Spiracles are located in the pleural area. Variation of this ground plan includes the fusion of terga or terga and sterna to form continuous dorsal or ventral shields or a conical tube. Some insects bear a sclerite in the pleural area called a laterotergite. Ventral sclerites are sometimes called laterosternites. During the embryonic stage of many insects and the postembryonic stage of primitive insects, 11 abdominal segments are present. In modern insects there is a tendency toward reduction in the number of the abdominal segments, but the primitive number of 11 is maintained during embryogenesis. Variation in abdominal segment number is considerable. If the Apterygota are considered to be indicative of the ground plan for pterygotes, confusion reigns: adult Protura have 12 segments, Collembola have 6. The orthopteran family Acrididae has 11 segments, and a fossil specimen of Zoraptera has a 10-segmented abdomen.

Exoskeleton

The insect outer skeleton, the cuticle, is made up of two layers: the epicuticle, which is a thin and waxy water resistant outer layer and contains no chitin, and a lower layer called the procuticle. The procuticle is chitinous and much thicker than the epicuticle and has two layers: an outer layer known as the exocuticle and an inner layer known as the endocuticle. The tough and flexible endocuticle is built from numerous layers of fibrous chitin and proteins, criss-crossing each other in a sandwich pattern, while the exocuticle is rigid and hardened. The exocuticle is greatly reduced in many soft-bodied insects (e.g., caterpillars), especially during their larval stages.

Insects are the only invertebrates to have developed active flight capability, and this has played an important role in their success. Their muscles are able to contract multiple times for each single nerve impulse, allowing the wings to beat faster than would ordinarily be possible. Having their muscles attached to their exoskeletons is more efficient and allows more muscle connections; crustaceans also use the same method, though all spiders use hydraulic pressure to extend their legs, a system inherited from their pre-arthropod ancestors. Unlike insects, though, most aquatic crustaceans are biomineralized with calcium carbonate extracted from the water.

Internal

Nervous System

The nervous system of an insect can be divided into a brain and a ventral nerve cord. The head capsule is made up of six fused segments, each with either a pair of ganglia, or a cluster of nerve cells outside of the brain. The first three pairs of ganglia are fused into the brain, while the three following pairs are fused into a structure of three pairs of ganglia under the insect's esophagus, called the subesophageal ganglion.

The thoracic segments have one ganglion on each side, which are connected into a pair, one pair per segment. This arrangement is also seen in the abdomen but only in the first eight segments. Many species of insects have reduced numbers of ganglia due to fusion or reduction. Some cockroaches have just six ganglia in the abdomen, whereas the wasp *Vespa crabro* has only two in the thorax and three in the abdomen. Some insects, like the house fly *Musca domestica*, have all the body ganglia fused into a single large thoracic ganglion.

At least a few insects have nociceptors, cells that detect and transmit signals responsible for the sensation of pain. This was discovered in 2003 by studying the variation in reactions of larvae of the common fruitfly Drosophila to the touch of a heated probe and an unheated one. The larvae reacted to the touch of the heated probe with a stereotypical rolling behavior that was not exhibited when the larvae were touched by the unheated probe. Although nociception has been demonstrated in insects, there is no consensus that insects feel pain consciously

Insects are capable of learning.

Digestive System

An insect uses its digestive system to extract nutrients and other substances from the food it consumes. Most of this food is ingested in the form of macromolecules and other complex substances like proteins, polysaccharides, fats and nucleic acids. These macromolecules must be broken down by catabolic reactions into smaller molecules like amino acids and simple sugars before being used by cells of the body for energy, growth, or reproduction. This break-down process is known as digestion.

The main structure of an insect's digestive system is a long enclosed tube called the alimentary canal, which runs lengthwise through the body. The alimentary canal directs food unidirectionally from the mouth to the anus. It has three sections, each of which performs a different process of digestion. In addition to the alimentary canal, insects also have paired salivary glands and salivary reservoirs. These structures usually reside in the thorax, adjacent to the foregut.

The salivary glands (element 30 in numbered diagram) in an insect's mouth produce saliva. The salivary ducts lead from the glands to the reservoirs and then forward through the head to an opening called the salivarium, located behind the hypopharynx. By moving its mouthparts (element 32 in numbered diagram) the insect can mix its food with saliva. The mixture of saliva and food then travels through the salivary tubes into the mouth, where it begins to break down. Some insects, like flies, have extra-oral digestion. Insects using extra-oral digestion expel digestive enzymes onto their food to break it down. This strategy allows insects to extract a significant proportion of the available nutrients from the food source.[31] The gut is where almost all of insects' digestion takes place. It can be divided into the foregut, midgut and hindgut.

Foregut

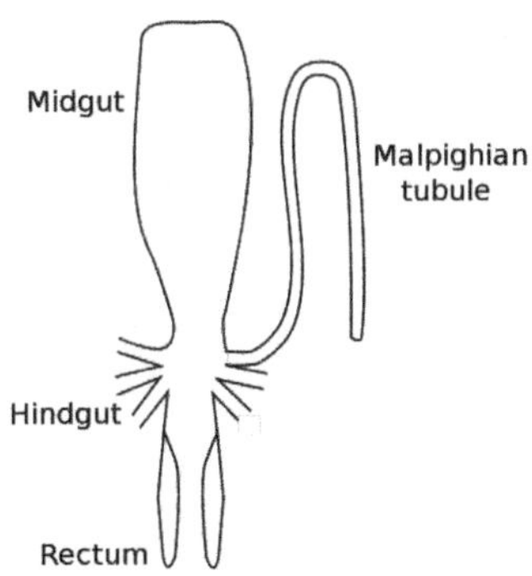

Stylized diagram of insect digestive tract showing malpighian tubule, from an insect of the order Orthoptera.

The first section of the alimentary canal is the foregut (element 27 in numbered diagram), or stomodaeum. The foregut is lined with a cuticular lining made of chitin and proteins as protection from tough food. The foregut includes the buccal cavity (mouth), pharynx, esophagus and crop and proventriculus (any part may be highly modified) which both store food and signify when to continue passing onward to the midgut.

Digestion starts in buccal cavity (mouth) as partially chewed food is broken down by saliva from the salivary glands. As the salivary glands produce fluid and carbohydrate-digesting enzymes (mostly amylases), strong muscles in the pharynx pump fluid into the buccal cavity, lubricating the food like the salivarium does, and helping blood feeders, and xylem and phloem feeders.

From there, the pharynx passes food to the esophagus, which could be just a simple tube passing it on to the crop and proventriculus, and then onward to the midgut, as in most insects. Alternately, the foregut may expand into a very enlarged crop and proventriculus, or the crop could just be a diverticulum, or fluid-filled structure, as in some Diptera species.[:30–31]

Bumblebee defecating. Note the contraction of the abdomen to provide internal pressure.

Midgut

Once food leaves the crop, it passes to the midgut (element 13 in numbered diagram), also known as the mesenteron, where the majority of digestion takes place. Microscopic projections from the midgut wall, called microvilli, increase the surface area of the wall and allow more nutrients to be absorbed; they tend to be close to the origin of the midgut. In some insects, the role of the microvilli and where they are located may vary. For example, specialized microvilli producing digestive enzymes may more likely be near the end of the midgut, and absorption near the origin or beginning of the midgut.

Hindgut

In the hindgut (element 16 in numbered diagram), or proctodaeum, undigested food particles are joined by uric acid to form fecal pellets. The rectum absorbs 90% of the water in these fecal pellets, and the dry pellet is then eliminated through the anus (element 17), completing the process of digestion. The uric acid is formed using hemolymph waste products diffused from the Malpighian tubules (element 20). It is then emptied directly into the alimentary canal, at the junction between the midgut and hindgut. The number of Malpighian tubules possessed by a given insect varies between species, ranging from only two tubules in some insects to over 100 tubules in others.

Reproductive System

The reproductive system of female insects consist of a pair of ovaries, accessory glands, one or more spermathecae, and ducts connecting these parts. The ovaries are made up of a number of egg tubes, called ovarioles, which vary in size and number by species. The number of eggs that the insect is able to make vary by the number of ovarioles with the rate that eggs can be develop being also influenced by ovariole design. Female insects are able make eggs, receive and store sperm, manipulate sperm from different males, and lay eggs. Accessory glands or glandular parts of the oviducts produce a variety of substances for sperm maintenance, transport and fertilization, as well as for protection of eggs. They can produce glue and protective substances for coating eggs or tough coverings for a batch of eggs called oothecae. Spermathecae are tubes or sacs in which sperm can be stored between the time of mating and the time an egg is fertilized.

For males, the reproductive system is the testis, suspended in the body cavity by tracheae and the fat body. Most male insects have a pair of testes, inside of which are sperm tubes or follicles that are enclosed within a membranous sac. The follicles connect to the vas deferens by the vas efferens, and the two tubular vasa deferentia connect to a median ejaculatory duct that leads to the outside. A portion of the vas deferens is often enlarged to form the seminal vesicle, which stores the sperm before they are discharged into the female. The seminal vesicles have glandular linings that secrete nutrients for nourishment and maintenance of the sperm. The ejaculatory duct is derived from an invagination of the epidermal cells during development and, as a result, has a cuticular lining. The terminal portion of the ejaculatory duct may be sclerotized to form the intromittent organ, the aedeagus. The remainder of the male reproductive system is derived from embryonic mesoderm, except for the germ cells, or spermatogonia, which descend from the primordial pole cells very early during embryogenesis.

Respiratory system

The tube-like heart (green) of the mosquito *Anopheles gambiae* extends horizontally across the body, interlinked with the diamond-shaped wing muscles (also green) and surrounded by pericardial cells (red). Blue depicts cell nuclei.

Insect respiration is accomplished without lungs. Instead, the insect respiratory system uses a system of internal tubes and sacs through which gases either diffuse or are actively pumped, delivering oxygen directly to tissues that need it via their trachea (element 8 in numbered diagram). Since oxygen is delivered directly, the circulatory system is not used to carry oxygen, and is therefore greatly reduced. The insect circulatory system has no veins or arteries, and instead consists of little more than a single, perforated dorsal tube which pulses peristaltically. Toward the thorax, the dorsal tube (element 14) divides into chambers and acts like the insect's heart. The opposite end of the dorsal tube is like the aorta of the insect circulating the hemolymph, arthropods' fluid analog of blood, inside the body cavity. Air is taken in through openings on the sides of the abdomen called spiracles.

The respiratory system is an important factor that limits the size of insects. As insects get bigger, this type of oxygen transport gets less efficient and thus the heaviest insect currently weighs less than 100 g. However, with increased atmospheric oxygen levels, as happened in the late Paleozoic, larger insects were possible, such as dragonflies with wingspans of more than two feet.

There are many different patterns of gas exchange demonstrated by different groups of insects. Gas exchange patterns in insects can range from continuous and diffusive ventilation, to discontinuous gas exchange. During continuous gas exchange, oxygen is taken in and carbon dioxide is released in a continuous cycle. In discontinuous gas exchange, however, the insect takes in oxygen while it is active and small amounts of carbon dioxide are released when the insect is at rest.Diffusive ventilation is simply a form of continuous gas exchange that occurs by diffusion rather than physically taking in the oxygen. Some species of insect that are submerged also have adaptations to aid in respiration. As larvae, many insects have gills that can extract oxygen dissolved in water, while others need to rise to the water surface to replenish air supplies which may be held or trapped in special structures.

Circulatory System

The insect circulatory system utilizes hemolymph, a tissue analogous to blood that circulates in the interior of the insect body, while remaining in direct contact with the animal's tissues. It is composed of plasma in which hemocytes are suspended. In addition to hemocytes, the plasma also contains many chemicals. It is also the major tissue type of the open circulatory system of arthropods, characteristic of spiders, crustaceans and insects.

REPRODUCTION AND DEVELOPMENT

The majority of insects hatch from eggs. The fertilization and development takes place inside the egg, enclosed by a shell (chorion) that consists of maternal tissue. In contrast to eggs of other arthropods, most insect eggs are drought resistant. This is because inside the chorion two additional membranes develop from embryonic tissue, the amnion and the serosa. This serosa secretes a cuticle rich in chitin that protects the embryo against desiccation. In Schizophora however the serosa does not develop, but these flies lay their eggs in damp places, such as rotting matter. Some species of insects, like the cockroach *Blaptica dubia*, as well as juvenile aphids and tsetse flies, are ovoviviparous. The eggs of ovoviviparous animals develop entirely inside the female, and then hatch immediately upon being laid. Some other species, such as those in the genus of cockroaches known as *Diploptera*, are viviparous, and thus gestate inside the mother and are born alive. Some insects, like parasitic wasps, show polyembryony, where a single fertilized egg divides into many and in some cases thousands of separate embryos. Insects may be *univoltine*, *bivoltine* or *multivoltine*, i.e. they may have one, two or many broods (generations) in a year.

A pair of *Simosyrphus grandicornis* hoverflies mating in flight.

A pair of *grasshoppers* mating.

The different forms of the male (top) and female (bottom) tussock moth*Orgyia recens* is an example of sexual dimorphism in insects.

Other developmental and reproductive variations include haplodiploidy, polymorphism, paedomorphosis or peramorphosis, sexual dimorphism, parthenogenesis and more rarely hermaphroditism.[:143] In haplodiploidy, which is a type of sex-determination system, the offspring's sex is determined by the number of sets of chromosomes an individual receives. This system is typical in bees and wasps. Polymorphism is where a species may have different *morphs* or *forms*, as in the oblong winged katydid, which has four different varieties: green, pink and yellow or tan. Some insects may retain phenotypes that are normally only seen in juveniles; this is called paedomorphosis. In peramorphosis, an opposite sort of phenomenon, insects take on previously unseen traits after they have matured into adults. Many insects display sexual dimorphism, in which males and females have notably different appearances, such as the moth *Orgyia recens* as an exemplar of sexual dimorphism in insects.

Some insects use parthenogenesis, a process in which the female can reproduce and give birth without having the eggs fertilized by a male. Many aphids undergo a form of parthenogenesis, called cyclical parthenogenesis, in which they alternate between one or many generations of asexual and sexual reproduction. In summer, aphids are generally female and parthenogenetic; in the autumn, males may be produced for sexual reproduction. Other insects produced by parthenogenesis are bees, wasps and ants, in which they spawn males. However, overall, most individuals are female, which are produced by fertilization. The males are haploid and the females are diploid. More rarely, some insects display hermaphroditism, in which a given individual has both male and female reproductive organs.

Insect life-histories show adaptations to withstand cold and dry conditions. Some temperate region insects are capable of activity during winter, while some others migrate to a warmer climate or go into a state of torpor. Still other insects have evolved mechanisms of diapause that allow eggs or pupae to survive these conditions.

Metamorphosis

Metamorphosis in insects is the biological process of development all insects must undergo. There are two forms of metamorphosis: incomplete metamorphosis and complete metamorphosis.

Incomplete Metamorphosis

Hemimetabolous insects, those with incomplete metamorphosis, change gradually by undergoing a series of molts. An insect molts when it outgrows its exoskeleton, which does not stretch and would otherwise restrict the insect's growth. The molting process begins as the insect's epidermis secretes a new epicuticle inside the old one. After this new epicuticle is secreted, the epidermis releases a mixture of enzymes that digests the endocuticle and thus detaches the old cuticle. When this stage is complete, the insect makes its body swell by taking in a large quantity of water or air, which makes the old cuticle split along predefined weaknesses where the old exocuticle was thinnest.[:142]

Immature insects that go through incomplete metamorphosis are called nymphs or in the case of dragonflies and damselflies, also naiads. Nymphs are similar in form to the adult except for the presence of wings, which are not developed until adulthood. With each molt, nymphs grow larger and become more similar in appearance to adult insects.

This Southern Hawker dragonfly molts its exoskeleton several times during its life as a nymph; shown is the final molt to become a winged adult (eclosion).

Complete Metamorphosis

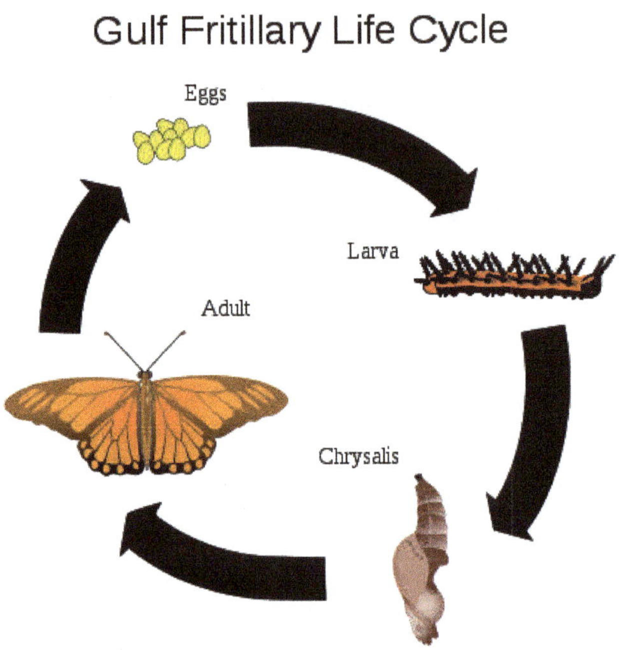

Gulf Fritillary life cycle, an example of holometabolism.

Holometabolism, or complete metamorphosis, is where the insect changes in four stages, an egg or embryo, a larva, a pupa and the adult or imago. In these species, an egg hatches to produce a larva, which is generally worm-like in form. This worm-like form can be one of several varieties: eruciform (caterpillar-like), scarabaeiform (grub-like), campodeiform (elongated, flattened and active), elateriform (wireworm-like) or vermiform (maggot-like). The larva grows and eventually becomes a pupa, a stage marked by reduced movement and often sealed within a cocoon. There are three types of pupae: obtect, exarate or coarctate. Obtect pupae are compact, with the legs and other appendages enclosed. Exarate pupae have their legs and other appendages free and extended. Coarctate pupae develop inside the larval skin.[:151] Insects undergo considerable change in form during the pupal stage, and emerge as adults. Butterflies are a well-known example of insects that undergo complete metamorphosis, although most insects use this life cycle. Some insects have evolved this system to hypermetamorphosis.

Some of the oldest and most successful insect groups, such Endopterygota, use a system of complete metamorphosis. Complete metamorphosis is unique to a group of certain insect orders including Diptera, Lepidoptera and Hymenoptera. This form of development is exclusive and not seen in any other arthropods.

SENSES AND COMMUNICATION

Many insects possess very sensitive and, or specialized organs of perception. Some insects such as bees can perceive ultraviolet wavelengths, or detect polarized light, while the antennae of male moths can detect the pheromones of female moths over distances of many kilometers. The yellow paper wasp (*Polistes versicolor*) is known for its wagging movements as a form of communication within the colony; it can waggle with a frequency of 10.6±2.1 Hz (n=190). These wagging movements can signal the arrival of new material into the nest and aggression between workers can be used to stimulate others to increase foraging expeditions. There is a pronounced tendency for there to be a trade-off between visual acuity and chemical or tactile acuity, such that most insects with well-developed eyes have reduced or simple antennae, and vice versa. There are a variety of different mechanisms by which insects perceive sound, while the patterns are not universal, insects can generally hear sound if they can produce it. Different insect species can have varying hearing, though most insects can hear only a narrow range of frequencies related to the frequency of the sounds they can produce. Mosquitoes have been found to hear up to 2 kHz, and some grasshoppers can hear up to 50 kHz. Certain predatory and parasitic insects can detect the characteristic sounds made by their prey or hosts, respectively. For instance, some nocturnal moths can perceive the ultrasonic emissions of bats, which helps them avoid predation. Insects that feed on blood have special sensory structures that can detect infrared emissions, and use them to home in on their hosts.

Some insects display a rudimentary sense of numbers, such as the solitary wasps that prey upon a single species. The mother wasp lays her eggs in individual cells and provides each egg with a number of live caterpillars on which the young feed when hatched. Some species of wasp always provide five, others twelve, and others as high as twenty-four caterpillars per cell. The number of caterpillars is different among species, but always the same for each sex of larva. The male solitary wasp in the genus *Eumenes* is smaller than the female, so the mother of one species supplies him with only five caterpillars; the larger female receives ten caterpillars in her cell.

Light Production and Vision

A few insects, such as members of the families Poduridae and Onychiuridae (Collembola), Mycetophilidae (Diptera) and the beetle families Lampyridae, Phengodidae, Elateridae and Staphylinidae are bioluminescent. The most familiar group are the fireflies, beetles of the family Lampyridae. Some species are able to control this light generation to produce flashes. The function varies with some species using them to attract mates, while others use them to lure prey. Cave dwelling larvae of *Arachnocampa* (Mycetophilidae, Fungus gnats) glow to lure small flying insects into sticky strands of silk. Some fireflies of the genus *Photuris* mimic the flashing of female *Photinus* species to attract males of that species, which are then captured and devoured. The colors of emitted light vary from dull blue (*Orfelia fultoni*, Mycetophilidae) to the familiar greens and the rare reds (*Phrixothrix tiemanni*, Phengodidae).

Most insects have compound eyes and two antennae.

Most insects, except some species of cave crickets, are able to perceive light and dark. Many species have acute vision capable of detecting minute movements. The eyes may include simple eyes or ocelli as well as compound eyes of varying sizes. Many species are able to detect light in the infrared, ultraviolet and the visible light wavelengths. Color vision has been demonstrated in many species and phylogenetic analysis suggests that UV-green-blue trichromacy existed from at least the Devonian period between 416 and 359 million years ago.

Sound production and hearing

Insects were the earliest organisms to produce and sense sounds. Insects make sounds mostly by mechanical action of appendages. In grasshoppers and crickets, this is achieved by stridulation. Cicadas make the loudest sounds among the insects by producing and amplifying sounds with special modifications to their body and musculature. The African cicada*Brevisana brevis* has been measured at 106.7 decibels at a distance of 50 cm (20 in). Some insects, such as the *Helicoverpa zea* moths, hawk moths and Hedylid butterflies, can hear ultrasound and take evasive action when they sense that they have been detected by bats. Some moths produce ultrasonic clicks that were once thought to have a role in jamming bat echolocation. The ultrasonic clicks were subsequently found to be produced mostly by unpalatable moths to warn bats, just as warning colorations are used against predators that hunt by sight. Some otherwise palatable moths have evolved to mimic these calls. More recently, the claim that some moths can jam bat sonar has been revisited. Ultrasonic recording and high-speed infrared videography of bat-moth interactions suggest the palatable tiger moth really does defend against attacking big brown bats using ultrasonic clicks that jam bat sonar.

Very low sounds are also produced in various species of Coleoptera, Hymenoptera, Lepidoptera, Mantodea and Neuroptera. These low sounds are simply the sounds made by the insect's movement. Through microscopic stridulatory structures located on the insect's muscles and joints, the normal sounds of the insect moving are amplified and can be used to warn or communicate with other insects. Most sound-making insects also have tympanal organs that can perceive airborne sounds. Some species in Hemiptera, such as the corixids (water boatmen), are known to communicate via underwater sounds. Most insects are also able to sense vibrations transmitted through surfaces.

Communication using surface-borne vibrational signals is more widespread among insects because of size constraints in producing air-borne sounds. Insects cannot effectively produce low-frequency sounds, and high-frequency sounds tend to disperse more in a dense environment (such as foliage), so insects living in such environments communicate primarily using substrate-borne vibrations. The mechanisms of production of vibrational signals are just as diverse as those for producing sound in insects.

Some species use vibrations for communicating within members of the same species, such as to attract mates as in the songs of the shield bug *Nezara viridula*. Vibrations can also be used to communicate between entirely different species; lycaenid (gossamer-winged butterfly) caterpillars which are myrmecophilous (living in a mutualistic association with ants) communicate with ants in this way. The Madagascar hissing cockroach has the ability to press air through its spiracles to make a hissing noise as a sign of aggression; the Death's-head Hawkmoth makes a squeaking noise by forcing air out of their pharynx when agitated, which may also reduce aggressive worker honey bee behavior when the two are in close proximity.

Chemical Communication

Chemical communications in animals rely on a variety of aspects including taste and smell. Chemoreception is the physiological response of a sense organ (i.e. taste or smell) to a chemical stimulus where the chemicals act as signals to regulate the state or activity of a cell. A semiochemical is a message-carrying chemical that is meant to attract, repel, and convey information. Types of semiochemicals include pheromones and kairomones. One example is the butterfly *Phengaris arion* which uses chemical signals as a form of mimicry to aid in predation.

In addition to the use of sound for communication, a wide range of insects have evolved chemical means for communication. These chemicals, termed semiochemicals, are often derived from plant metabolites include those meant to attract, repel and provide other kinds of information. Pheromones, a type of semiochemical, are used for attracting mates of the opposite sex, for aggregating conspecific individuals of both sexes, for deterring other individuals from approaching, to mark a trail, and to trigger aggression in nearby individuals. Allomonea benefit their producer by the effect they have upon the receiver. Kairomones benefit their receiver instead of their producer. Synomones benefit the producer and the receiver. While some chemicals are targeted at individuals of the same species, others are used for communication across species. The use of scents is especially well known to have developed in social insects.

SOCIAL BEHAVIOR

A cathedral mound created by termites (Isoptera).

Social insects, such as termites, ants and many bees and wasps, are the most familiar species of eusocial animal. They live together in large well-organized colonies that may be so tightly integrated and genetically similar that the colonies of some species are sometimes considered superorganisms. It is sometimes argued that the various species of honey bee are the only invertebrates (and indeed one of the few non-human groups) to have evolved a system of abstract symbolic communication where a behavior is used to *represent* and convey specific information about something in the environment. In this communication system, called dance language, the angle at which a bee dances represents a direction relative to the sun, and the length of the dance represents the distance to be flown.[:309–311] Though perhaps not as advanced as honey bees, bumblebees also potentially have some social communication behaviors. *Bombus terrestris*, for example, exhibit a faster learning curve for visiting unfamiliar, yet rewarding flowers, when they can see a conspecific foraging on the same species.

Only insects which live in nests or colonies demonstrate any true capacity for fine-scale spatial orientation or homing. This can allow an insect to return unerringly to a single hole a few millimeters in diameter among thousands of apparently identical holes clustered together, after a trip of up to several kilometers' distance. In a phenomenon known as philopatry, insects that hibernate have shown the ability to recall a specific location up to a year after last viewing the area of interest. A few insects seasonally migrate large distances between different geographic regions (e.g., the overwintering areas of the Monarch butterfly).

Care of Young

The eusocial insects build nest, guard eggs, and provide food for offspring full-time (see Eusociality). Most insects, however, lead short lives as adults, and rarely interact with one another except to mate or compete for mates. A small number exhibit some form of parental care, where they will at least guard their eggs, and sometimes continue guarding their offspring until adulthood, and possibly even feeding them. Another simple form of parental care is to construct a nest (a burrow or an actual construction, either of which may be simple or complex), store provisions in it, and lay an egg upon those provisions. The adult does not contact the growing offspring, but it nonetheless does provide food. This sort of care is typical for most species of bees and various types of wasps.

LOCOMOTION

Flight

White-lined sphinx moth feeding in flight

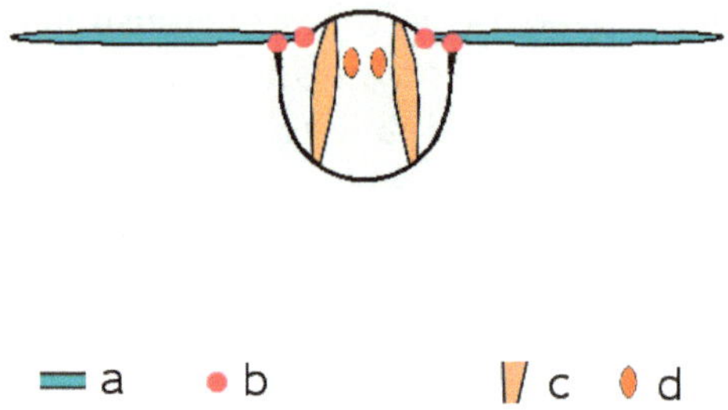

Basic motion of the insect wing in insect with an indirect flight mechanism scheme of dorsoventral cut through a thorax segment with..

a wings

b joints

c dorsoventral muscles

d longitudinal muscles.

Insects are the only group of invertebrates to have developed flight. The evolution of insect wings has been a subject of debate. Some entomologists suggest that the wings are from paranotal lobes, or extensions from the insect's exoskeleton called the nota, called the *paranotal theory*. Other theories are based on a pleural origin. These theories include suggestions that wings originated from modified gills, spiracular flaps or as from an appendage of the epicoxa. The *epicoxal theory* suggests the insect wings are modified epicoxal exites, a modified appendage at the base of the legs or coxa. In the Carboniferous age, some of the *Meganeura* dragonflies had as much as a 50 cm (20 in) wide wingspan. The appearance of gigantic insects has been found to be consistent with high atmospheric oxygen. The respiratory system of insects constrains their size, however the high oxygen in the atmosphere allowed larger sizes. The largest flying insects today are much smaller and include several moth species such as the Atlas moth and the White Witch (*Thysania agrippina*).

Insect flight has been a topic of great interest in aerodynamics due partly to the inability of steady-state theories to explain the lift generated by the tiny wings of insects. But insect wings are in motion, with flapping and vibrations, resulting in churning and eddies, and the misconception that physics says "bumblebees can't fly" persisted throughout most of the twentieth century.

Unlike birds, many small insects are swept along by the prevailing windsalthough many of the larger insects are known to make migrations. Aphids are known to be transported long distances by low-level jet streams. As such, fine line patterns associated with converging winds within weather radar imagery, like the WSR-88D radar network, often represent large groups of insects.

Walking

Many adult insects use six legs for walking and have adopted a tripedal gait. The tripedal gait allows for rapid walking while always having a stable stance and has been studied extensively in cockroaches. The legs are used in alternate triangles touching the ground. For the first step, the middle right leg and the front and rear left legs are in contact with the ground and move the insect forward, while the front and rear right leg and the middle left leg are lifted and moved forward to a new position. When they touch the ground to form a new stable triangle the other legs can be lifted and brought forward in turn and so on. The purest form of the tripedal gait is seen in insects moving at high speeds. However, this type of locomotion is not rigid and insects can adapt a variety of gaits. For example, when moving slowly, turning, or avoiding obstacles, four or more feet may be touching the ground. Insects can also adapt their gait to cope with the loss of one or more limbs.

Cockroaches are among the fastest insect runners and, at full speed, adopt a bipedal run to reach a high velocity in proportion to their body size. As cockroaches move very quickly, they need to be video recorded at several hundred frames per second to reveal their gait. More sedate locomotion is seen in the stick insects or walking sticks (Phasmatodea). A few insects have evolved to walk on the surface of the water, especially members of the Gerridae family, commonly known as water striders. A few species of ocean-skaters in the genus *Halobates* even live on the surface of open oceans, a habitat that has few insect species.

Use in robotics

Insect walking is of particular interest as an alternative form of locomotion in robots. The study of insects and bipeds has a significant impact on possible robotic methods of transport. This may allow new robots to be designed that can traverse terrain that robots with wheels may be unable to handle.

Swimming

A large number of insects live either part or the whole of their lives underwater. In many of the more primitive orders of insect, the immature stages are spent in an aquatic environment. Some groups of insects, like certain water beetles, have aquatic adults as well.

Many of these species have adaptations to help in under-water locomotion. Water beetles and water bugs have legs adapted into paddle-like structures. Dragonfly naiads use jet propulsion, forcibly expelling water out of their rectal chamber. Some species like the water striders are capable of walking on the surface of water. They can do this because their claws are not at the tips of the legs as in most insects, but recessed in a special groove further up the leg; this prevents the claws from piercing the water's surface film. Other insects such as the Rove beetle *Stenus* are known to emit pygidial gland secretions that reduce surface tension making it possible for them to move on the surface of water by Marangoni propulsion (also known by the German term *Entspannungsschwimmen*).

The backswimmer *Notonecta glauca* underwater, showing its paddle-like hindleg adaptation.

ECOLOGY

Insect ecology is the scientific study of how insects, individually or as a community, interact with the surrounding environment or ecosystem. Insects play one of the most important roles in their ecosystems, which includes many roles, such as soil turning and aeration, dung burial, pest control, pollination and wildlife nutrition. An example is the beetles, which are scavengers that feed on dead animals and fallen trees and thereby recycle biological materials into forms found useful by other organisms. These insects, and others, are responsible for much of the process by which topsoil is created.

Defense and Predation

Insects are mostly soft bodied, fragile and almost defenseless compared to other, larger lifeforms. The immature stages are small, move slowly or are immobile, and so all stages are exposed to predation and parasitism. Insects then have a variety of defense strategies to avoid being attacked by predators or parasitoids. These include camouflage, mimicry, toxicity and active defense.

Camouflage is an important defense strategy, which involves the use of coloration or shape to blend into the surrounding environment. This sort of protective coloration is common and widespread among beetle families, especially those that feed on wood or vegetation, such as many of the leaf beetles (family Chrysomelidae) or weevils. In some of these species, sculpturing or various colored scales or hairs cause the beetle to resemble bird dung or other inedible objects. Many of those that live in sandy environments blend in with the coloration of the substrate. Most phasmids are known for effectively replicating the forms of sticks and leaves, and the bodies of some species (such as *O. macklotti* and *Palophus centaurus*) are covered in mossy or lichenous outgrowths that supplement their disguise. Some species have the ability to change color as their surroundings shift (*B. scabrinota*, *T. californica*). In a further behavioral adaptation to supplement crypsis, a number of species have been noted to perform a rocking motion where the body is swayed from side to side that is thought to reflect the movement of leaves or twigs swaying in the breeze. Another method by which stick insects avoid predation and resemble twigs is by feigning death (catalepsy), where the insect enters a motionless state that can be maintained for a long period. The nocturnal feeding habits of adults also aids Phasmatodea in remaining concealed from predators.

Perhaps one of the most well-known examples of mimicry, the viceroy butterfly (top) appears very similar to the noxious-tasting monarch butterfly (bottom).

Another defense that often uses color or shape to deceive potential enemies is mimicry. A number of longhorn beetles (family Cerambycidae) bear a striking resemblance to wasps, which helps them avoid predation even though the beetles are in fact harmless. Batesian and Müllerian mimicry complexes are commonly found in Lepidoptera. Genetic polymorphism and natural selection give rise to otherwise edible species (the mimic) gaining a survival advantage by resembling inedible species (the model). Such a mimicry complex is referred to as *Batesian* and is most commonly known by the mimicry by the limenitidine Viceroy butterfly of the inedible danaine Monarch. Later research has discovered that the Viceroy is, in fact more toxic than the Monarch and this resemblance should be considered as a case of Müllerian mimicry. In Müllerian mimicry, inedible species, usually within a taxonomic order, find it advantageous to resemble each other so as to reduce the sampling rate by predators who need to learn about the insects' inedibility. Taxa from the toxic genus *Heliconius* form one of the most well known Müllerian complexes.

Chemical defense is another important defense found amongst species of Coleoptera and Lepidoptera, usually being advertised by bright colors, such as the Monarch butterfly. They obtain their toxicity by sequestering the chemicals from the plants they eat into their own tissues. Some Lepidoptera manufacture their own toxins. Predators that eat poisonous butterflies and moths may become sick and vomit violently, learning not to eat those types of species; this is actually the basis of Müllerian mimicry. A predator who has previously eaten a poisonous lepidopteran may avoid other species with similar markings in the future, thus saving many other species as well. Some ground beetles of the Carabidae family can spray chemicals from their abdomen with great accuracy, to repel predators.

Pollination

European honey bee carrying pollen in a pollen basket back to the hive.

Pollination is the process by which pollen is transferred in the reproduction of plants, thereby enabling fertilisation and sexual reproduction. Most flowering plants require an animal to do the transportation. While other animals are included as pollinators, the majority of pollination is done by insects. Because insects usually receive benefit for the pollination in the form of energy rich nectar it is a grand example of mutualism. The various flower traits (and combinations thereof) that differentially attract one type of pollinator or another are known as pollination syndromes. These arose through complex plant-animal adaptations. Pollinators find flowers through bright colorations, including ultraviolet, and attractant pheromones. The study of pollination by insects is known as *anthecology*.

Parasitism

Many insects are parasites of other insects such as the parasitoid wasps. These insects are known as entomophagous parasites. They can be beneficial due to their devastation of pests that can destroy crops and other resources. Many insects have a parasitic relationship with humans such as the mosquito. These insects are known to spread diseases such as malaria and yellow fever and because of such, mosquitoes indirectly cause more deaths of humans than any other animal.

RELATIONSHIP TO HUMANS

As pests

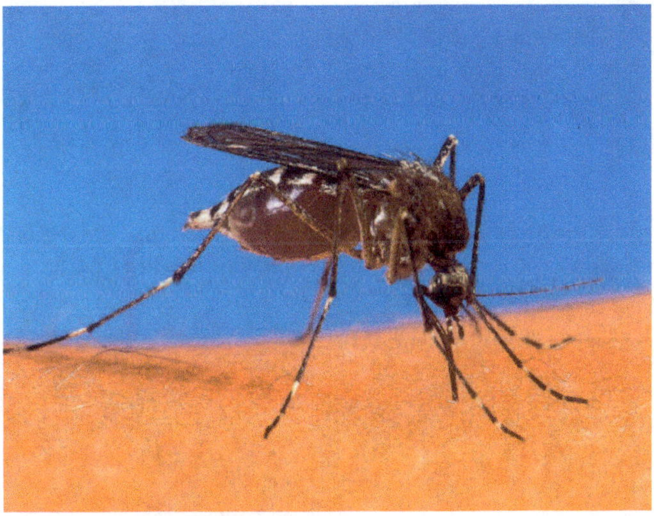

Aedes aegypti, a parasite, is the vector of dengue fever and yellow fever.

Many insects are considered pests by humans. Insects commonly regarded as pests include those that are parasitic (*e.g.* lice, bed bugs), transmit diseases (mosquitoes, flies), damage structures (termites), or destroy agricultural goods (locusts, weevils). Many entomologists are involved in various forms of pest control, as in research for companies to produce insecticides, but increasingly rely on methods of biological pest control, or biocontrol. Biocontrol uses one organism to reduce the population density of another organism — the pest — and is considered a key element of integrated pest management.

Despite the large amount of effort focused at controlling insects, human attempts to kill pests with insecticides can backfire. If used carelessly, the poison can kill all kinds of organisms in the area, including insects' natural predators, such as birds, mice and other insectivores. The effects of DDT's use exemplifies how some insecticides can threaten wildlife beyond intended populations of pest insects.

In Beneficial Roles

Because they help flowering plants to cross-pollinate, some insects are critical to agriculture. This European honey bee is gathering nectar while pollen collects on its body.

A robberfly with its prey, a hoverfly. Insectivorous relationships such as these help control insect populations.

Although pest insects attract the most attention, many insects are beneficial to the environment and to humans. Some insects, like wasps, bees, butterflies and ants, pollinate flowering plants. Pollination is a mutualistic relationship between plants and insects. As insects gather nectar from different plants of the same species, they also spread pollen from plants on which they have previously fed. This greatly increases plants' ability to cross-pollinate, which maintains and possibly even improves their evolutionary fitness. This ultimately affects humans since ensuring healthy crops is critical to agriculture. As well as pollination ants help with seed distribution of plants. This helps to spread the plants which increases plant diversity. This leads to an overall better environment. A serious environmental problem is the decline of populations of pollinator insects, and a number of species of insects are now cultured primarily for pollination management in order to have sufficient pollinators in the field, orchard or greenhouse at bloom time.[:240–243] Another solution, as shown in Delaware, has been to raise native plants to help support native pollinators like *L. vierecki*.Insects also produce useful substances such as honey, wax, lacquer and silk. Honey bees have been cultured by humans for thousands of years for honey, although contracting for crop pollination is becoming more significant for beekeepers. The silkworm has greatly affected human history, as silk-driven trade established relationships between China and the rest of the world.

Insectivorous insects, or insects which feed on other insects, are beneficial to humans because they eat insects that could cause damage to agriculture and human structures. For example, aphids feed on crops and cause problems for farmers, but ladybugs feed on aphids, and can be used as a means to get significantly reduce pest aphid populations. While birds are perhaps more visible predators of insects, insects themselves account for the vast majority of insect consumption. Ants also help control animal populations by consuming small vertebrates. Without predators to keep them in check, insects can undergo almost unstoppable population explosions.

Insects are also used in medicine, for example fly larvae (maggots) were formerly used to treat wounds to prevent or stop gangrene, as they would only consume dead flesh. This treatment is finding modern usage in some hospitals. Recently insects have also gained attention as potential sources of drugs and other medicinal substances. Adult insects, such as crickets and insect larvae of various kinds, are also commonly used as fishing bait.

In Research

Insects play important roles in biological research. For example, because of its small size, short generation time and high fecundity, the common fruit fly *Drosophila melanogaster* is a model organism for studies in the genetics of higher eukaryotes. *D. melanogaster* has been an essential part of studies into principles like genetic linkage, interactions between genes, chromosomal genetics, development, behavior and evolution. Because genetic systems are well conserved among eukaryotes, understanding basic cellular processes like DNA replication or transcription in fruit flies can help to understand those processes in other eukaryotes, including humans.The genome of *D. melanogaster* was sequenced in 2000, reflecting the organism's important role in biological research. It was found that 70% of the fly genome is similar to the human genome, supporting the evolution theory.

The common fruitfly *Drosophila melanogaster* is one of the most widely used organisms in biological research.

As Food

In some cultures, insects, especially deep-fried cicadas, are considered to be delicacies, whereas in other places they form part of the normal diet. Insects have a high protein content for their mass, and some authors suggest their potential as a major source of protein in human nutrition. In most first-world countries, however, entomophagy (the eating of insects), is taboo. Since it is impossible to entirely eliminate pest insects from the human food chain, insects are inadvertently present in many foods, especially grains. Food safety laws in many countries do not prohibit insect parts in food, but rather limit their quantity. According to cultural materialist anthropologist Marvin Harris, the eating of insects is taboo in cultures that have other protein sources such as fish or livestock.

Due to the abundance of insects and a worldwide concern of food shortages, the Food and Agriculture Organisation of the United Nations considers that the world may have to, in the future, regard the prospects of eating insects as a food staple. Insects are noted for their nutrients, having a high content of protein, minerals and fats and are eaten by one-third of the global population.

In Culture

Scarab beetles held religious and cultural symbolism in Old Egypt, Greece and some shamanistic Old World cultures. The ancient Chinese regarded cicadas as symbols of rebirth or immortality. In Mesopotamian literature, the epic poem of Gilgamesh has allusions to Odonata which signify the impossibility of immortality. Amongst the Aborigines of Australia of the Arrernte language groups, honey ants and witchety grubs served as personal clan totems. In the case of the 'San' bush-men of the Kalahari, it is the praying mantis which holds much cultural significance including creation and zen-like patience in waiting.

REFERENCES

1. Engel, Michael S.; David A. Grimaldi (2004). "New light shed on the oldest insect". Nature. 427 (6975): 627–630. Bibcode:2004Natur.427..627E. doi:10.1038/nature02291. PMID 14961119.
2. Chapman, A. D. (2006). Numbers of living species in Australia and the World. Canberra: Australian Biological Resources Study. ISBN 978-0-642-56850-2.
3. Wilson, E.O. "Threats to Global Diversity". Archived from the original on 20 February 2015. Retrieved 17 May 2009.
4. Novotny, Vojtech; Basset, Yves; Miller, Scott E.; Weiblen, George D.; Bremer, Birgitta; Cizek, Lukas; Drozd, Pavel (2002). "Low host specificity of herbivorous insects in a tropical forest". Nature. 416 (6883): 841–844. Bibcode:2002Natur.416..841N. doi:10.1038/416841a. PMID 11976681.
5. Erwin, Terry L. (1997). Biodiversity at its utmost: Tropical Forest Beetles. pp. 27–40. In: Reaka-Kudla, M. L.; Wilson, D. E.; Wilson, E. O. (eds.). Biodiversity II. Joseph Henry Press, Washington, D.C.
6. Erwin, Terry L. (1982). "Tropical forests: their richness in Coleoptera and other arthropod species". Coleopt. Bull. 36: 74–75.
7. "insect physiology" McGraw-Hill Encyclopedia of Science and Technology, Ch. 9, p. 233, 2007
8. Vincent Brian Wigglesworth. "Insect". Encyclopædia Britannica online. Retrieved 19 April 2012.
9. Harper, Douglas; Dan McCormack (November 2001). "Online Etymological Dictionary". Online Etymological Dictionary. LogoBee.com. p. 1. Retrieved 1 November 2011.
10. Russell Garwood; Gregory Edgecombe (2011). "Early terrestrial animals, evolution and uncertainty". Evolution, Education, and Outreach. 4 (3): 489–501. doi:10.1007/s12052-011-0357-y.
11. "Palaeos invertebrates:Arthropoda". Palaeos Invertebrates. 3 May 2002. Archived from the original on 15 February 2009. Retrieved 6 May 2009.
12. Misof, et.al., Bernhard (7 November 2014). "Phylogenomics resolves the timing and pattern of insect evolution". Science. 346 (6210): 763–767. Bibcode:2014Sci...346..763M. doi:10.1126/science.1257570. PMID 25378627.
13. "Evolution of insect flight". Malcolm W. Browne. 25 October 1994. Retrieved 6 May 2009.
14. "Tree of Life Web Project. Version 1 January 1995 (temporary) of Arthropoda". Tree of Life Web Project. 1995. Retrieved 9 May 2009.
15. "Researchers Discover Oldest Fossil Impression of a Flying Insect". Newswise. 14 October 2008. Retrieved 21 September 2014.
16. Rice, C. M.; Ashcroft, W. A.; Batten, D. J.; Boyce, A. J.; Caulfield, J. B. D.; Fallick, A. E.; Hole, M. J.; Jones, E.; Pearson, M. J.; Rogers, G.; Saxton, J. M.; Stuart, F. M.; Trewin, N. H.; Turner, G. (1995). "A Devonian auriferous hot

spring system, Rhynie, Scotland". Journal of the Geological Society, London. 152 (2): 229–250. doi:10.1144/gsjgs.152.2.0229.

17. Wiegmann BM, Trautwein MD, Winkler IS, Barr NB, Kim JW, Lambkin C, Bertone MA, Cassel BK, Bayless KM, Heimberg AM, Wheeler BM, Peterson KJ, Pape T, Sinclair BJ, Skevington JH, Blagoderov V, Caravas J, Kutty SN, Schmidt-Ott U, Kampmeier GE, Thompson FC, Grimaldi DA, Beckenbach AT, Courtney GW, Friedrich M, Meier R, Yeates DK (2011). "Episodic radiations in the fly tree of life". Proceedings of the National Academy of Sciences. 108 (14): 5690–5695. Bibcode:2011PNAS..108.5690W. doi:10.1073/pnas.1012675108. PMC 3078341Freely accessible. PMID 21402926.
18. Grimaldi, D.; Engel, M. S. (2005). Evolution of the Insects. Cambridge University Press. ISBN 0-521-82149-5.
19. Garwood, Russell J.; Sutton, Mark D. (2010). "X-ray micro-tomography of Carboniferous stem-Dictyoptera: New insights into early insects". Biology Letters. 6 (5): 699–702. doi:10.1098/rsbl.2010.0199. PMC 2936155Freely accessible. PMID 20392720. Retrieved 9 June 2015.
20. Rasnitsyn, A. P.; Quicke, D. L. J. (2002). History of Insects. Kluwer Academic Publishers. ISBN 1-4020-0026-X.
21. J. Stein Carter (29 March 2005). "Coevolution and Pollination". University of Cincinnati. Archived from the original on 30 April 2009. Retrieved 9 May 2009.
22. Sahney, S.; Benton, M. J.; Falcon-Lang, H. J. (2010). "Rainforest collapse triggered Pennsylvanian tetrapod diversification in Euramerica". Geology. 38 (12): 1079–1082. Bibcode:2010Geo....38.1079S. doi:10.1130/G31182.1.
23. "Coevolution and Pollination". University of Cincinnati. Archived from the original on 30 April 2009. Retrieved 9 May 2009.
24. "Insecta". Tree of Life Web Project. 2002. Retrieved 12 May 2009.
25. Gullan, P.J.; Cranston, P.S. (2005). The Insects: An Outline of Entomology (3 ed.). Oxford: Blackwell Publishing. ISBN 1-4051-1113-5.
26. Kendall, David A. (2009). "Classification of Insect". Retrieved 9 May 2009.
27. Gilliott, Cedric (1995). Entomology (2 ed.). Springer-Verlag New York, LLC. p. 96. ISBN 0-306-44967-6.
28. Kapoor, V.C. C. (1998). Principles and Practices of Animal Taxonomy. 1 (1 ed.). Science Publishers. p. 48. ISBN 1-57808-024-X.
29. Johnson, K. P.; Yoshizawa, K.; Smith, V. S. (2004). "Multiple origins of parasitism in lice". Proceedings of the Royal Society of London. 271 (1550): 1771–1776. doi:10.1098/rspb.2004.2798. PMC 1691793Freely accessible. PMID 15315891.
30. Terry, M. D.; Whiting, M. F. (2005). "Mantophasmatodea and phylogeny of the lower neopterous insects". Cladistics. 21 (3): 240–257. doi:10.1111/j.1096-0031.2005.00062.x.
31. Lo, Nathan; Tokuda, Gaku; Watanabe, Hirofumi; Rose, Harley; Slaytor, Michael; Maekawa, Kiyoto; Bandi, Claudio; Noda, Hiroaki (2000). "Evidence from multiple gene sequences indicates that termites evolved from wood-feeding cockroaches".

Current Biology. 10 (13): 801–804. doi:10.1016/S0960-9822(00)00561-3. PMID 10898984.

32. Bonneton, F.; Brunet, F. G.; Kathirithamby J. & Laudet, V. (2006). "The rapid divergence of the ecdysone receptor is a synapomorphy for Mecopterida that clarifies the Strepsiptera problem". Insect Molecular Biology. 15 (3): 351–362. doi:10.1111/j.1365-2583.2006.00654.x. PMID 16756554.
33. Whiting, M.F. (2002). "Mecoptera is paraphyletic: multiple genes and phylogeny of Mecoptera and Siphonaptera". Zoologica Scripta. 31 (1): 93–104. doi:10.1046/j.0300-3256.2001.00095.x.
34. Stork, N. E.; McBroom, J; Gely, C; Hamilton, A. J. (2015). "New approaches narrow global species estimates for beetles, insects, and terrestrial arthropods". Proceedings of the National Academy of Sciences. 112 (24): 7519–23. doi:10.1073/pnas.1502408112. PMC 4475949Freely accessible. PMID 26034274.
35. "O. Orkin Insect zoo". The University of Nebraska Department of Entomology. Archived from the original on 2 June 2009. Retrieved 3 May 2009.
36. Resh, Vincent H.; Ring T. Carde (2009). Encyclopedia of Insects (2 ed.). U. S. A.: Academic Press. ISBN 0-12-374144-0.
37. Barnes, R.S.K.; Calow, P.; Olive, P.; Golding, D.; and Spicer, J. (2001). "Invertebrates with Legs: the Arthropods and Similar Groups". The Invertebrates: A Synthesis. Blackwell Publishing. p. 168. ISBN 0-632-04761-5.
38. Lowenstam, H.A.; Weiner, S. (1989). On biomineralization. Oxford University Press US. p. 111. ISBN 0-19-504977-2.
39. Schneiderman, Howard A. (1960). "Discontinuous respiration in insects: role of the spiracles". Biol. Bull. 119 (3): 494–528. doi:10.2307/1539265. JSTOR 1539265.
40. Eisemann, C. H.; Jorgensen, W. K.; Merritt, D. J.; Rice, M. J.; Cribb, B. W.; Webb, P. D.; Zalucki, M. P. (1984). "Do insects feel pain? — A biological view". Cellular and Molecular Life Sciences. 40 (2): 1420–1423. doi:10.1007/BF01963580.
41. Tracey, J; Wilson, RI; Laurent, G; Benzer, S (2003). "painless, a Drosophila gene essential for nociception". Cell. 113 (2): 261–273. doi:10.1016/S0092-8674(03)00272-1. PMID 12705873.
42. Sømme, LS (14 January 2005). "Sentience and pain in invertebrates" (PDF). Norwegian Scientific Committee for Food Safety. Retrieved 30 September 2009.
43. Annual Review of Entomology, Evolutionary Biology of Insect Learning, Vol. 53: 145-160
44. "General Entomology – Digestive and Excritory system". NC state University. Retrieved 3 May 2009.
45. "General Entomology – Digestive and Excretory system". NC state University. Retrieved 3 May 2009.
46. Duncan, Carl D. (1939). A Contribution to The Biology of North American

Vespine Wasps (1 ed.). Stanford: Stanford University Press. pp. 24–29.

47. Nation, James L. (2001). "Digestion". Insect Physiology and Biochemistry (1st ed.). CRC Press. ISBN 0-8493-1181-0.
48. Meyer, John R. (17 February 2006). "Circulatory System". NC State University: Department of Entomology, NC State University. p. 1. Retrieved 11 October 2009.
49. "What Keeps Bugs from Being Bigger?". Argonne National Laboratory. 8 August 2007. Retrieved 15 July 2013.
50. Chown, S.L.; S.W. Nicholson (2004). Insect Physiological Ecology. New York: Oxford University Press. ISBN 0-19-851549-9.
51. Richard W. Merritt; Kenneth W. Cummins; Martin B. Berg (editors) (2007). An Introduction to the Aquatic Insects of North America (4th ed.). Kendall Hunt Publishers. ISBN 978-0-7575-5049-2.
52. Merritt, RW; KW Cummins & MB Berg (2007). An Introduction To The Aquatic Insects Of North America. Kendall Hunt Publishing Company. ISBN 0-7575-4128-3.
53. Chapman, R.F. (1998). The Insects; Structure and Function (4th ed.). Cambridge, UK: Cambridge University Press. ISBN 0521578906.
54. Wyatt, G. R. (1961). "The Biochemistry of Insect Hemolymph". Annual Review of Entomology. 6: 75–102. doi:10.1146/annurev.en.06.010161.000451.
55. Jacobs, C.G.; Rezende, G.L.; Lamers, G.E.; van der Zee, M. (2013). "The extraembryonic serosa protects the insect egg against desiccation". Proceedings of the Royal Society – Biological series. 280 (1764): 20131082. doi:10.1098/rspb.2013.1082. PMID 23782888.
56. "Glossary of Lepidopteran and Odonate anatomy". Rare species atlas. Virginia Department of Conservation and Recreation. 2013. Retrieved 14 June 2013.
57. Hughes, William O. H.; Oldroyd, Benjamin P.; Beekman, Madeleine; Ratnieks, Francis L. W. (2008). "Ancestral Monogamy Shows Kin Selection Is Key to the Evolution of Eusociality". Science. American Association for the Advancement of Science. 320 (5880): 1213–1216. Bibcode:2008Sci...320.1213H. doi:10.1126/science.1156108. PMID 18511689.
58. Nevo, E.; Coll, M. (2001). "Effect of nitrogen fertilization on Aphis gossypii (Homoptera: Aphididae): variation in size, color, and reproduction". J. Econ. Entomol. 94 (1): 27–32. doi:10.1603/0022-0493-94.1.27. PMID 11233124.
59. Jahn, G. C.; Almazan, L. P.; Pacia, J. (2005). "Effect of nitrogen fertilizer on the intrinsic rate of increase of the rusty plum aphid, Hysteroneura setariae (Thomas) (Homoptera: Aphididae) on rice (Oryza sativa L.)" (PDF). Environmental Entomology. 34 (4): 938–943. doi:10.1603/0046-225X-34.4.938.
60. Debbie Hadley. "Where do insects go in winter?". About.com. Retrieved 19 April 2012.
61. Lee, Richard E Jr. (1989). "Insect Cold-Hardiness: To Freeze or Not to Freeze"

(PDF). BioScience. 39 (5): 308–313. doi:10.2307/1311113. JSTOR 1311113.

62. Ruppert, E.E.; Fox, R.S. & Barnes, R.D. (2004). Invertebrate Zoology (7 ed.). Brooks / Cole. pp. 523–524. ISBN 0-03-025982-7.
63. "Insects" (PDF). Alien Life Forms. p. 4. Retrieved 17 May 2009.
64. Esch, Harald (1971). "Wagging Movements in the Wasp Polistes versicolor Vulgaris Bequaert". Zeitschrift für Vergleichende Physiologie. 72 (3): 221–25. doi:10.1007/bf00297781.
65. Cator, L. J.; Arthur, BJ; Harrington, LC; Hoy, RR (2009). "Harmonic convergence in the love songs of the dengue vector mosquito". Science. 323 (5917): 1077–1079. Bibcode:2009Sci...323.1077C. doi:10.1126/science.1166541. PMC 2847473Freely accessible. PMID 19131593.
66. Möller, R. (2002). A Biorobotics Approach to the Study of Insect Visual Homing Strategies (PDF) (in German). p. 11.
67. Pugsley, Chris W. (1983). "Literature review of the New Zealand glowworm Arachnocampa luminosa (Diptera: Keroplatidae) and related cave-dwelling Diptera" (PDF). New Zealand Entomologist. 7 (4): 419–424. doi:10.1080/00779962.1983.9722435.
68. Lloyd, James E. (1984). "Occurrence of Aggressive Mimicry in Fireflies". The Florida Entomologist. 67 (3): 368–376. doi:10.2307/3494715. JSTOR 3494715.
69. Lloyd, James E.; Erin C. Gentry (2003). The Encyclopedia of Insects. Academic Press. pp. 115–120. ISBN 0-12-586990-8
70. Briscoe, AD; Chittka, L (2001). "The evolution of color vision in insects". Annu. Rev. Entomol. 46: 471–510. doi:10.1146/annurev.ento.46.1.471. PMID 11112177.
71. Walker, T.J., ed. (2001). University of Florida Book of Insect Records.
72. Kay, Robert E. (1969). "Acoustic signalling and its possible relationship to assembling and navigation in the moth, Heliothis zea". Journal of Insect Physiology. 15 (6): 989–1001. doi:10.1016/0022-1910(69)90139-5.
73. Spangler, Hayward G. (1988). "Moth hearing, defense, and communication". Annual Review of Entomology. 33 (1): 59–81. doi:10.1146/annurev.ento.33.1.59.
74. Hristov, N.I.; Conner, W.E. (2005). "Sound strategy: acoustic aposematism in the bat–tiger moth arms race". Naturwissenschaften. 92 (4): 164–169. Bibcode:2005NW.....92..164H. doi:10.1007/s00114-005-0611-7. PMID 15772807.
75. Barber, J. R.; W. E. Conner (2007). "Acoustic mimicry in a predator–prey interaction". Proc. Natl. Acad. Sci. 104 (22): 9331–9334. Bibcode:2007PNAS..104.9331B. doi:10.1073/pnas.0703627104. PMC 1890494Freely accessible. PMID 17517637.
76. Corcoran, Aaron J.; Jesse R. Barber; William E. Conner (2009). "Tiger Moth Jams Bat Sonar". Science. 325 (5938): 325–327. Bibcode:2009Sci...325..325C. doi:10.1126/science.1174096.
77. Theiss, Joachim (1982). "Generation and radiation of sound by stridulating water insects as exemplified by the corixids". Behavioral Ecology and Sociobiology. 10 (3): 225–235. doi:10.1007/BF00299689.

78. Virant-Doberlet, M.; Čokl A. (2004). "Vibrational communication in insects". Neotropical Entomology. 33 (2): 121–134. doi:10.1590/S1519-566X2004000200001.

79. Bennet-Clark, H.C. (1998). "Size and scale effects as constraints in insect sound communication". Phil. Trans. R. Soc. Lond. B. 353 (1367): 407–419. doi:10.1098/rstb.1998.0219.

80. Miklas, Nadège; Stritih, Nataša; Čokl, Andrej; Virant-Doberlet, Meta; Renou, Michel (2001). "The Influence of Substrate on Male Responsiveness to the Female Calling Song in Nezara viridula". Journal of Insect Behavior. 14 (3): 313–332. doi:10.1023/A:1011115111592.

81. DeVries, P. J. (1990). "Enhancement of symbiosis between butterfly caterpillars and ants by vibrational communication". Science. 248 (4959): 1104–1106. Bibcode:1990Sci...248.1104D. doi:10.1126/science.248.4959.1104. PMID 17733373.

82. Nelson, Margaret C.; Jean Fraser (1980). "Sound production in the cockroach, Gromphadorhina portentosa: evidence for communication by hissing". Behavioral Ecology and Sociobiology. 6 (4): 305–314. doi:10.1007/BF00292773.

83. Moritz, R. F. A.; Kirchner, W. H.; Crewe, R. M. (1991). "Chemical camouflage of the death's head hawkmoth (Acherontia atropos L.) in honeybee colonies". Naturwissenschaften. 78 (4): 179–182. Bibcode:1991NW.....78..179M. doi:10.1007/BF01136209.

84. Thomas, Jeremy; Schönrogge, Karsten; Bonelli, Simona; Barbero, Francesca; Balletto, Emilio (2010). "Corruption of ant acoustical signals by mimetic social parasites". Communicative and Integrative Biology. 3 (2): 169–171. doi:10.4161/cib.3.2.10603. PMC 2889977Freely accessible. PMID 20585513.

85. Brewer, Gary. "Social insects". North Dakota State University. Archived from the original on 21 March 2008. Retrieved 6 May 2009.

86. Leadbeater, E.; L. Chittka (2007). "The dynamics of social learning in an insect model, the bumblebee (Bombus terrestris)". Behavioral Ecology and Sociobiology. 61 (11): 1789–1796. doi:10.1007/s00265-007-0412-4.

87. Salt, R.W. (1961). "Principles of Insect Cold-Hardiness". Annual Review of Entomology. Lethbirdge, Alberta, Canada. 6: 55–74. doi:10.1146/annurev.en.06.010161.000415.

88. "Social Insects". North Dakota State University. Archived from the original on 21 March 2008. Retrieved 12 October 2009.

89. Jockusch, EL; Ober, KA (September 2004). "Hypothesis testing in evolutionary developmental biology: a case study from insect wings". Journal of Heredity. 95 (5): 382–396. doi:10.1093/jhered/esh064. PMID 15388766.

90. Dudley, R (1998). "Atmospheric oxygen, giant Paleozoic insects and the evolution of aerial locomotor performance" (PDF). Journal of Experimental Biology. 201 (8): 1043–1050.

91. Yates, Diana (2008). Birds migrate together at night in dispersed flocks, new study indicates. University of Illinois at Urbana – Champaign. Retrieved on 26 April 2009.

92. Drake, V. A.; R. A. Farrow (1988). "The Influence of Atmospheric Structure and Motions on Insect Migration". Annual Review of Entomology. 33: 183–210. doi:10.1146/annurev.en.33.010188.001151.

93. Bart Geerts and Dave Leon (2003). P5A.6 Fine-Scale Vertical Structure of a Cold Front As Revealed By Airborne 95 GHZ Radar. University of Wyoming. Retrieved on 26 April 2009.

94. Biewener, Andrew A (2003). Animal Locomotion. Oxford University Press. ISBN 0-19-850022-X.

95. Ikawa, Terumi; Okabe, Hidehiko; Hoshizaki, Sugihiko; Kamikado, Takahiro; Cheng, Lanna (2004). "Distribution of the oceanic insects Halobates (Hemiptera: Gerridae) off the south coast of Japan". Entomological Science. 7 (4): 351–357. doi:10.1111/j.1479-8298.2004.00083.x.

96. Mill, P. J.; R. S. Pickard (1975). "Jet-propulsion in anisopteran dragonfly larvae". Journal of Comparative Physiology A. 97 (4): 329–338. doi:10.1007/BF00631969.

97. Linsenmair, K.; Jander R. (1976). "Das "entspannungsschwimmen" von Velia and Stenus". Naturwissenschaften. 50 (6): 231. Bibcode:1963NW.....50..231L. doi:10.1007/BF00639292.

98. Bush, J. W. M.; David L. Hu (2006). "Walking on Water: Biolocomotion at the Interface" (PDF). Annu. Rev. Fluid Mech. 38: 339–369. Bibcode:2006AnRFM..38..339B. doi:10.1146/annurev.fluid.38.050304.092157. Archived from the original (PDF) on 10 July 2007.

99. Schowalter, Timothy Duane (2006). Insect ecology: an ecosystem approach (2(illustrated) ed.). Academic Press. p. 572. ISBN 978-0-12-088772-9.

100. Losey, John E.; Vaughan, Mace (2006). "The Economic Value of Ecological Services Provided by Insects". BioScience. 56 (4): 311–323(13). doi:10.1641/0006-3568(2006)56[311:TEVOES]2.0.CO;2.

101. Ritland, D.; L. P. Brower (1991). "The viceroy butterfly is not a Batesian mimic". Nature. 350 (6318): 497–498. Bibcode:1991Natur.350..497R. doi:10.1038/350497a0. Viceroys are as unpalatable as monarchs, and significantly more unpalatable than queens from representative Florida populations.

102. Evans, Arthur V.; Charles Bellamy (2000). An Inordinate Fondness for Beetles. University of California Press. ISBN 978-0-520-22323-3.

103. "Photos: Masters of Disguise—Amazing Insect Camouflage". Retrieved 11 June 2015.

104. Bedford, Geoffrey O. (1978). "Biology and Ecology of the Phasmatodea". Annual Review of Entomology. 23: 125–149. doi:10.1146/annurev.en.23.010178.001013.

105. Ritland, D. B.; L. P. Brower (1991). "The viceroy butterfly is not a Batesian mimic". Nature. 350 (6318): 497–498. Bibcode:1991Natur.350..497R.

doi:10.1038/350497a0. Viceroys are as unpalatable as monarchs, and significantly more unpalatable than queens from representative Florida populations

106. Meyer, A. (2006). "Repeating Patterns of Mimicry". PLoS Biology. 4 (10): e341. doi:10.1371/journal.pbio.0040341. PMC 1617347Freely accessible. PMID 17048984.
107. Kricher, John (1999). "6". A Neotropical Companion. Princeton University Press. pp. 157–158. ISBN 978-0-691-00974-2.
108. "Pollinator Factsheet" (PDF). United States Forest Service. Archived from the original (PDF) on 10 April 2008. Retrieved 19 April 2012.
109. Bale, JS; van Lenteren, JC; Bigler, F. (27 February 2008). "Biological control and sustainable food production". Philosophical Transactions of the Royal Society of London. Series B: Biological Sciences. 363 (1492): 761–776. doi:10.1098/rstb.2007.2182. PMC 2610108Freely accessible. PMID 17827110.
110. Davidson, E. (2006). Big Fleas Have Little Fleas: How Discoveries of Invertebrate Diseases Are Advancing Modern Science. Tucson, Ariz.: University of Arizona Press. ISBN 0-8165-2544-7.
111. Colborn, T; vom Saal, FS; Soto, AM (October 1993). "Developmental effects of endocrine-disrupting chemicals in wildlife and humans". Environmental Health Perspectives. 101 (5): 378–384. doi:10.2307/3431890. JSTOR 3431890. PMC 1519860Freely accessible. PMID 8080506.
112. Nakamaru, M; Iwasab, Y; Nakanishic, J (October 2003). "Extinction risk to bird populations caused by DDT exposure". Chemosphere. 53 (4): 377–387. doi:10.1016/S0045-6535(03)00010-9. PMID 12946395.
113. Holldobler, Wilson (1994). Journey to the ants: a story of scientific exploration. Westminster college McGill Library: Cambridge, Mass.:Belknap Press of Haravard University Press, 1994. pp. 196–199. ISBN 0-674-48525-4.
114. Smith, Deborah T (1991). Agriculture and the Environment: The 1991 Yearbook of Agriculture (1991 ed.). United States Government Printing. ISBN 0-16-034144-2.
115. Kuehn, F. Coordinator. (2015). "Farming for native bees". World Wide Web electronic publication. (Accessed: September 22, 2015).
116. Camargo, Rafael; Paulo Oliveira (2011). "Natural history of the Neotrobical arboreal ant, Odontomachus hastatus: Nest sites, foraging schedule, and diet". Journal of Insect Science. 12 (18): 1–9. doi:10.1673/031.012.4801. PMC 3476954Freely accessible. PMID 22957686.
117. "Biocontrol Network – Beneficial Insects". Biocontrol Network. Retrieved 9 May 2009.
118. Davidson, RH; William F. Lyon (1979). Insect Pests of Farm, Garden, and Orchard. Wiley, John & Sons. p. 38. ISBN 0-471-86314-9.
119. Dossey, Aaron T. (December 2010). "Insects and their chemical weaponry: New potential for drug discovery". Natural Product Reports. Royal Society of Chemistry (RSC Publishing). 27 (12): Pages 1737–1757. doi:10.1039/c005319h. PMID 20957283.

120. Sherman, Ronald A.; Pechter, Edward A. (1987). "Maggot therapy: a review of the therapeutic applications of fly larvae in human medicine, especially for treating osteomyelitis". Medical and Veterinary Entomology. 2 (3): Pages 225–230. doi:10.1111/j.1365-2915.1988.tb00188.x.

121. Pierce, BA (2006). Genetics: A Conceptual Approach (2nd ed.). New York: W.H. Freeman and Company. p. 87. ISBN 0-7167-8881-0.

122. Adams, MD; Celniker, SE; Holt, RA; Evans, CA; Gocayne, JD; Amanatides, PG; Scherer, SE; Li, PW; et al. (24 March 2000). "The genome sequence of Drosophila melanogaster". Science. 287 (5461): 2185–2195. Bibcode:2000Sci...287.2185.. doi:10.1126/science.287.5461.2185. PMID 10731132.

123. Michels, John (1880). John Michels, ed. Science. 1. American Association for the Advance of Science. 229 Broadway ave., N.Y.: American Association for the Advance of Science. pp. 2090pp. ISBN 1-930775-36-9.

124. Maierbrugger, Arno (14 May 2013). "UN: Insects are 'food of the future' (video)". Inside Investor. Retrieved 17 May 2013.

CHAPTER 2

Arachnology

Chapter Contents

- **Introduction**
- **Arachnology as a Science**
- **Arachnological Societies**
- **Popular Arachnology**
- **Spider**
- **Description**
- **Ecology and Behavior**
- **Scorpion**

Back in ancient Greece a maiden had the audacity to challenge the Goddess of Wisdom to a spinning contest. As a punishment she was for ever afterwards forced to spin and weave. Her name, ράχνη (Arachne) gave rise to the word used to study spiders and their relatives: Arachn[e] ology(science/study of).

Arachnids are an ancient group of creatures dating back to the Silurian period, more than 420 million years ago (earlier ones have allegedly been found but there status as Arachnids is arguable). They belong to the Arthropods ('jointed limbs') that account for over 80% of all living animals and include the insects, crustaceans, millipedes, and centipedes. Arthropods are characterised by their tough segmented outer shell, or exoskeleton, and jointed limbs.

INTRODUCTION

Arachnology is the scientific study of spiders and related animals such as scorpions, pseudoscorpions,and harvestmen, collectively called arachnids. Those who study spiders and other arachnids are **arachnologists**.

Nephila clavipes

ARACHNOLOGY AS A SCIENCE

Arachnologists are primarily responsible for classifying arachnids and studying aspects of their biology. In the popular imagination they are sometimes referred to as 'spider experts'. Disciplines within arachnology include naming species and determining their evolutionary relationships to one another (taxonomy and systematics), studying how they interact with other members of their species and/or their environment (behavioural ecology), or how they are distributed in different regions and habitats (faunistics). Other arachnologists perform research on the anatomy or physiology of arachnids, including the venom of spiders and scorpions. Others study the impact of spiders in agricultural ecosystems and whether they can be used as biological control agents.

Subdisciplines

Arachnology can be broken down into more specific specialties. These topics include:

- **acarology** - the study of ticks and mites.
- **araneology** - the study of spiders.
- **scorpiology** - the study of scorpions.

ARACHNOLOGICAL SOCIETIES

Arachnologists are served by a number of scientific societies, both national and international in scope. Their main role is to encourage the exchange of ideas between researchers, to organise meetings and congresses, and in a number of cases to publish academic journals. Some are also involved in science outreach programs, like the *European spider of the year*, which raise awareness of these animals among the general public.

International

- International Society of Arachnology (ISA) website

Africa

- African Arachnological Society (AFRAS) website

Asia

- Arachnological Society of Japan (ASJ) website
- Asian Society of Arachnology (ASA) website
- Indian Society of Arachnology website
- Iranian Arachnological Society (IAS) website

Australasia

- Australasian Arachnological Society website

Europe

- Aracnofilia - Associazione Italiana di Aracnologia website
- Arachnologia Belgica - Belgian Arachnological Society (ARABEL) website
- Arachnologische Gesellschaft (AraGes) website
- Association Francaise d'Arachnologie (AsFrA) website
- British Arachnological Society (BAS) website
- Czech Arachnological Society website
- European Society of Arachnology (ESA) website
- Grupo Ibérico de Aracnologia (Iberian Peninsula) website
- Magyar Arachnolgia - Hungarian Arachnology

North America

- American Arachnological Society (AAS) website

Arachnological Journals

Scientific journals devoted to the study of arachnids include:

- *Acarologia*
- *Acta Arachnologica* - published by the Arachnological Society of Japan
- *Arachnology* - published by the British Arachnological Society
- *Arachnology Letters* - published by the Arachnologische Gesellschaft
- *International Journal of Acarology*
- *Journal of Arachnology* - published by the American Arachnological Society
- *Revista Ibérica de Arachnología* - published by the Grupo Ibérico de Aracnología
- *Revue Arachnologique*

POPULAR ARACHNOLOGY

In the 1970s, arachnids - particularly tarantulas - started to become popular as exotic pets. Many tarantulas thus become more widely known by their common names such as the Mexican redknee tarantula (*Brachypelma smithi*).

Various societies now focus on the husbandry, care, study and captive breeding of tarantulas, and other arachnids. They also typically produce journals or newsletters with articles and advice on these subjects.

- British Tarantula Society (BTS) website
- Deutsche Arachnologische Gesellschaft (DeArGe) website
- The American Tarantula Society (ATS) website

SPIDER

Spiders (order **Araneae**) are air-breathing arthropods that have eight legs and chelicerae with fangs that inject venom. They are the largest order of arachnids and rank seventh in total species diversity among all other orders of organisms. Spiders are found worldwide on every continent except for Antarctica, and have become established in nearly every habitat with the exceptions of air and sea colonization. As of November 2015, at least 45,700 spider species, and 113 families have been recorded by taxonomists. However, there has been dissension within the scientific community as to how all these families should be classified, as evidenced by the over 20 different classifications that have been proposed since 1900.

Anatomically, spiders differ from other arthropods in that the usual body segments are fused into two tagmata, the cephalothorax and abdomen, and joined by a small, cylindrical pedicel. Unlike insects, spiders do not have antennae. In all except the most primitive group, the Mesothelae, spiders have the most centralized nervous systems of all arthropods, as all their ganglia are fused into one mass in the cephalothorax. Unlike most arthropods, spiders have no extensor muscles in their limbs and instead extend them by hydraulic pressure.

Their abdomens bear appendages that have been modified into spinnerets that extrude silk from up to six types of glands. Spider webs vary widely in size, shape and the amount of sticky thread used. It now appears that the spiral orb web may be one of the earliest forms, and spiders that produce tangled cobwebs are more abundant and diverse than orb-web spiders. Spider-like arachnids with silk-producing spigots appeared in the Devonian period about 386 million years ago, but these animals apparently lacked spinnerets. True spiders have been found in Carboniferous rocks from 318 to 299 million years ago, and are very similar to the most primitive surviving suborder, the Mesothelae. The main groups of modern spiders, Mygalomorphae and Araneomorphae, first appeared in the Triassic period, before 200 million years ago.

A herbivorous species, *Bagheera kiplingi*, was described in 2008, but all other known species are predators, mostly preying on insects and on other spiders, although a few large species also take birds and lizards. Spiders use a wide range of strategies to capture prey: trapping it in sticky webs, lassoing it with sticky bolas, mimicking the prey to avoid detection, or running it down. Most detect prey mainly by sensing vibrations, but the active hunters have acute vision, and hunters of the genus *Portia* show signs of intelligence in their choice of tactics and ability to develop new ones. Spiders' guts are too narrow to take solids, and they liquefy their food by flooding it with digestive enzymes. They also grind food with the bases of their pedipalps, as arachnids do not have the mandibles that crustaceans and insects have.

Male spiders identify themselves by a variety of complex courtship rituals to avoid being eaten by the females. Males of most species survive a few matings, limited mainly by their short life spans. Females weave silk egg-cases, each of which may contain hundreds of eggs. Females of many species care for their young, for example by carrying them around or by sharing food with them. A minority of species are social, building communal webs that may house anywhere from a few to 50,000 individuals. Social behavior ranges from precarious toleration, as in the widow spiders, to co-operative hunting and food-sharing. Although most spiders live for at most two years, tarantulas and other mygalomorph spiders can live up to 25 years in captivity.

While the venom of a few species is dangerous to humans, scientists are now researching the use of spider venom in medicine and as non-polluting pesticides. Spider silk provides a combination of lightness, strength and elasticity that is superior to that of synthetic materials, and spider silk genes have been inserted into mammals and plants to see if these can be used as silk factories. As a result of their wide range of behaviors, spiders have become common symbols in art and mythology symbolizing various combinations of patience, cruelty and creative powers. An abnormal fear of spiders is called arachnophobia.

DESCRIPTION

Body Plan

Palystes castaneus female dorsal aspect

1: pedipalp

2: trichobothria

3: carapace of prosoma (cephalothorax)

4: opisthosoma (abdomen)

5: eyes – AL (anterior lateral) AM (anterior median) PL (posterior lateral) PM (posterior median) Leg segments:

6: costa

7: trochanter

8: patella

9: tibia

10: metatarsus

11: tarsus

13: claw

14: chelicera

Nos 1 to 14 as for dorsal aspect

15: sternum of prosoma

16: pedicel (also called pedicle)

17: book lung sac

18: book lung stigma

19: epigastric fold

20: epigyne

21: anterior spinneret

22: posterior spinneret

I, II, III, IV = Leg numbers from anterior to posterior

Spiders are chelicerates and therefore arthropods. As arthropods they have: segmented bodies with jointed limbs, all covered in a cuticle made of chitin and proteins; heads that are composed of several segments that fuse during the development of the embryo. Being chelicerates, their bodies consist of two tagmata, sets of segments that serve similar functions: the foremost one, called the cephalothorax or prosoma, is a complete fusion of the segments that in an insect would form two separate tagmata, the head and thorax; the rear tagma is called the abdomen or opisthosoma. In spiders, the cephalothorax and abdomen are connected by a small cylindrical section, the pedicel. The pattern of segment fusion that forms chelicerates' heads is unique among arthropods, and what would normally be the first head segment disappears at an early stage of development, so that chelicerates lack the antennae typical of most arthropods. In fact, chelicerates' only appendages ahead of the mouth are a pair of chelicerae, and they lack anything that would function directly as "jaws". The first appendages behind the mouth are called pedipalps, and serve different functions within different groups of chelicerates.

Spiders and scorpions are members of one chelicerate group, the arachnids. Scorpions' chelicerae have three sections and are used in feeding. Spiders' chelicerae have two sections and terminate in fangs that are generally venomous, and fold away behind the upper sections while not in use. The upper sections generally have thick "beards" that filter solid lumps out of their food, as spiders can take only liquid food. Scorpions' pedipalps generally form large claws for capturing prey, while those of spiders are fairly small appendages whose bases also act as an extension of the mouth; in addition, those of male spiders have enlarged last sections used for sperm transfer.

In spiders, the cephalothorax and abdomen are joined by a small, cylindrical pedicel, which enables the abdomen to move independently when producing silk. The upper surface of the cephalothorax is covered by a single, convex carapace, while the underside is covered by two rather flat plates. The abdomen is soft and egg-shaped. It shows no sign of segmentation, except that the primitive Mesothelae, whose living members are the Liphistiidae, have segmented plates on the upper surface.

Circulation and Respiration

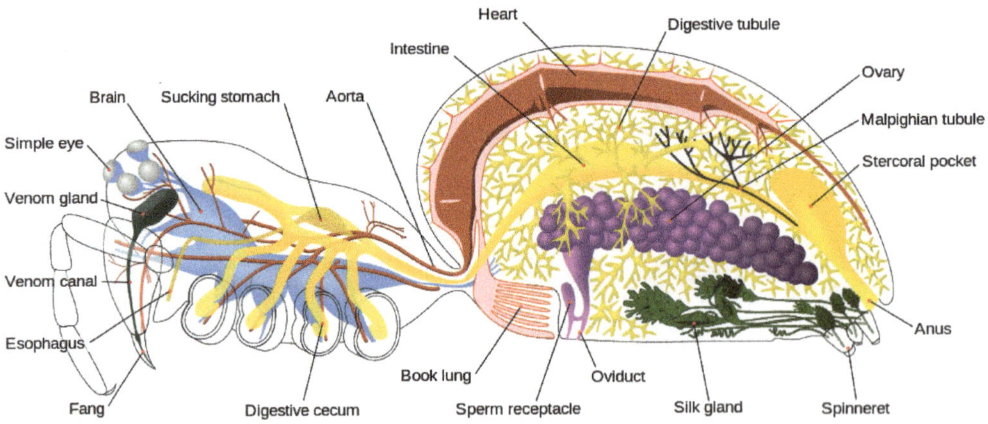

Like other arthropods, spiders are coelomates in which the coelom is reduced to small areas round the reproductive and excretory systems. Its place is largely taken by a hemocoel, a cavity that runs most of the length of the body and through which blood flows. The heart is a tube in the upper part of the body, with a few ostia that act as non-return valves allowing blood to enter the heart from the hemocoel but prevent it from leaving before it reaches the front end. However, in spiders, it occupies only the upper part of the abdomen, and blood is discharged into the hemocoel by one artery that opens at the rear end of the abdomen and by branching arteries that pass through the pedicle and open into several parts of the cephalothorax. Hence spiders have open circulatory systems. The blood of many spiders that have book lungs contains the respiratory pigmenthemocyanin to make oxygen transport more efficient.

Spiders have developed several different respiratory anatomies, based on book lungs, a tracheal system, or both. Mygalomorph and Mesothelae spiders have two pairs of book lungs filled with haemolymph, where openings on the ventral surface of the abdomen allow air to enter and diffuse oxygen. This is also the case for some basal araneomorph spiders, like the family Hypochilidae, but the remaining members of this group have just the anterior pair of book lungs intact while the posterior pair of breathing organs are partly or fully modified into tracheae, through which oxygen is diffused into the haemolymph or directly to the tissue and organs. The trachea system has most likely evolved in small ancestors to help resist desiccation. The trachea were originally connected to the surroundings through a pair of openings called spiracles, but in the majority of spiders this pair of spiracles has fused into a single one in the middle, and moved backwards close to the spinnerets. Spiders that have tracheae generally have higher metabolic rates and better water conservation. Spiders are ectotherms, so environmental temperatures affect their activity.

Feeding, Digestion and Excretion

Cheiracanthium punctorium, displaying fangs.

Uniquely among chelicerates, the final sections of spiders' chelicerae are fangs, and the great majority of spiders can use them to inject venom into prey from venom glands in the roots of the chelicerae. The family Uloboridae has lost its venom glands, and kills its prey with silk instead. Like most arachnids, including scorpions, spiders have a narrow gut that can only cope with liquid food and spiders have two sets of filters to keep solids out. They use one of two different systems of external digestion. Some pump digestive enzymes from the midgut into the prey and then suck the liquified tissues of the prey into the gut, eventually leaving behind the empty husk of the prey. Others grind the prey to pulp using the chelicerae and the bases of the pedipalps, while flooding it with enzymes; in these species, the chelicerae and the bases of the pedipalps form a preoral cavity that holds the food they are processing.

The stomach in the cephalothorax acts as a pump that sends the food deeper into the digestive system. The mid gut bears many digestive ceca, compartments with no other exit, that extract nutrients from the food; most are in the abdomen, which is dominated by the digestive system, but a few are found in the cephalothorax.

Most spiders convert nitrogenous waste products into uric acid, which can be excreted as a dry material. Malphigian tubules ("little tubes") extract these wastes from the blood in the hemocoel and dump them into the cloacal chamber, from which they are expelled through the anus.Production of uric acid and its removal via Malphigian tubules are a water-conserving feature that has evolved independently in several arthropod lineages that can live far away from water, for example the tubules of insects and arachnids develop from completely different parts of the embryo. However, a few primitive spiders, the sub-order Mesothelae and infra-order Mygalomorphae, retain the ancestral arthropod nephridia ("little kidneys"), which use large amounts of water to excrete nitrogenous waste products as ammonia.

Central Nervous System

The basic arthropod central nervous system consists of a pair of nerve cords running below the gut, with paired ganglia as local control centers in all segments; a brain formed by fusion of the ganglia for the head segments ahead of and behind the mouth, so that the esophagus is encircled by this conglomeration of ganglia. Except for the primitive Mesothelae, of which the Liphistiidae are the sole surviving family, spiders have the much more centralized nervous system that is typical of arachnids: *all* the ganglia of all segments behind the esophagus are fused, so that the cephalothorax is largely filled with nervous tissue and there are no ganglia in the abdomen; in the Mesothelae, the ganglia of the abdomen and the rear part of the cephalothorax remain unfused.

Despite the relatively small central nervous system, some spiders (like *Portia*) exhibit complex behaviour, including the ability to use a trial-and-error approach.

Sense Organs

Eyes

This jumping spider's main ocelli (center pair) are very acute. The outer pair are "secondary eyes" and there are other pairs of secondary eyes on the sides and top of its head.

Most spiders have four pairs of eyes on the top-front area of the cephalothorax, arranged in patterns that vary from one family to another. The pair at the front are of the type called pigment-cup ocelli ("little eyes"), which in most arthropods are only capable of detecting the direction from which light is coming, using the shadow cast by the walls of the cup. However, the main eyes at the front of spiders' heads are pigment-cup ocelli that are capable of forming images. The other eyes are thought to be derived from the compound eyes of the ancestral chelicerates, but no longer have the separate facets typical of compound eyes. Unlike the main eyes, in many spiders these secondary eyes detect light reflected from a reflective tapetum lucidum, and wolf spiders can be spotted by torch light reflected from the tapeta. On the other hand, jumping spiders' secondary eyes have no tapeta. Some jumping spiders' visual acuity exceeds by a factor of ten that of dragonflies, which have by far the best vision among insects; in fact the human eye is only about five times sharper than a jumping spider's. They achieve this by a telephoto-like series of lenses, a four-layer retina and the ability to swivel their eyes and integrate images from different stages in the scan. The downside is that the scanning and integrating processes are relatively slow.

There are spiders with a reduced number of eyes, of these those with six-eyes are the most numerous and are missing a pair of eyes on the anterior median line, others species have four-eyes and some just two. Cave dwelling species have no eyes, or possess vestigial eyes incapable of sight.

Other Senses

As with other arthropods, spiders' cuticles would block out information about the outside world, except that they are penetrated by many sensors or connections from sensors to the nervous system. In fact, spiders and other arthropods have modified their cuticles into elaborate arrays of sensors. Various touch sensors, mostly bristles called setae, respond to different levels of force, from strong contact to very weak air currents. Chemical sensors provide equivalents of taste and smell, often by means of setae. Pedipalps carry a large number of such setae sensitive to contact chemicals and air-borne smells, such as female pheromones. Spiders also have in the joints of their limbs slit sensillae that detect forces and vibrations. In web-building spiders, all these mechanical and chemical sensors are more important than the eyes, while the eyes are most important to spiders that hunt actively.

Like most arthropods, spiders lack balance and acceleration sensors and rely on their eyes to tell them which way is up. Arthropods' proprioceptors, sensors that report the force exerted by muscles and the degree of bending in the body and joints, are well understood. On the other hand, little is known about what other internal sensors spiders or other arthropods may have.

Locomotion

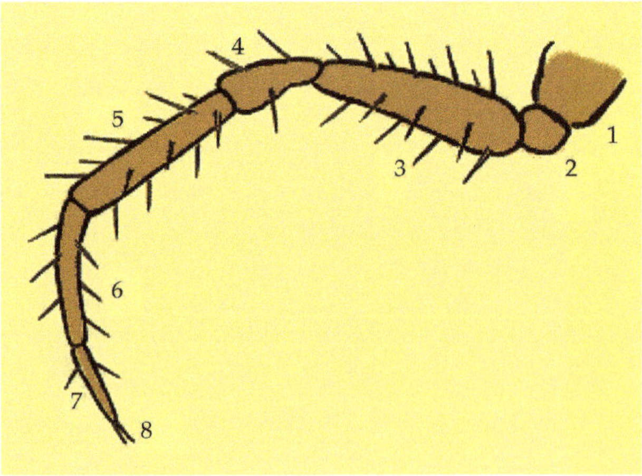

Image of a spider leg: 1–coxa; 2–trochanter; 3–femur; 4–patella; 5–tibia; 6–metatarsus; 7–tarsus; 8–claws.

Each of the eight legs of a spider consists of seven distinct parts. The part closest to and attaching the leg to the cephalothorax is the coxa; the next segment is the short trochanter that works as a hinge for the following long segment, the femur; next is the spider's knee, the patella, which acts as the hinge for the tibia; the metatarsus is next, and it connects the tibia to the tarsus (which may be thought of as a foot of sorts); the tarsus ends in a claw made up of either two or three points, depending on the family to which the spider belongs. Although all arthropods use muscles attached to the inside of the exoskeleton to flex their limbs, spiders and a few other groups still use hydraulic pressure to extend them, a system inherited from their pre-arthropod ancestors. The only extensor muscles in spider legs are located in the three hip joints (bordering the coxa and the trochanter). As a result, a spider with a punctured cephalothorax cannot extend its legs, and the legs of dead spiders curl up. Spiders can generate pressures up to eight times their resting level to extend their legs, and jumping spiders can jump up to 50 times their own length by suddenly increasing the blood pressure in the third or fourth pair of legs. Although larger spiders use hydraulics to straighten their legs, unlike smaller jumping spiders they depend on their flexor muscles to generate the propulsive force for their jumps.

Most spiders that hunt actively, rather than relying on webs, have dense tufts of fine hairs between the paired claws at the tips of their legs. These tufts, known as scopulae, consist of bristles whose ends are split into as many as 1,000 branches, and enable spiders with scopulae to walk up vertical glass and upside down on ceilings. It appears that scopulae get their grip from contact with extremely thin layers of water on surfaces.Spiders, like most other arachnids, keep at least four legs on the surface while walking or running.

Silk Production

An orb weaver producing silk from its spinnerets.

The abdomen has no appendages except those that have been modified to form one to four (usually three) pairs of short, movable spinnerets, which emit silk. Each spinneret has many spigots, each of which is connected to one silk gland. There are at least six types of silk gland, each producing a different type of silk.

Silk is mainly composed of a protein very similar to that used in insect silk. It is initially a liquid, and hardens not by exposure to air but as a result of being drawn out, which changes the internal structure of the protein. It is similar in tensile strength to nylon and biological materials such as chitin, collagen and cellulose, but is much more elastic. In other words, it can stretch much further before breaking or losing shape.

Some spiders have a cribellum, a modified spinneret with up to 40,000 spigots, each of which produces a single very fine fiber. The fibers are pulled out by the calamistrum, a comb-like set of bristles on the jointed tip of the cribellum, and combined into a composite woolly thread that is very effective in snagging the bristles of insects. The earliest spiders had cribella, which produced the first silk capable of capturing insects, before spiders developed silk coated with sticky droplets. However, most modern groups of spiders have lost the cribellum.

Tarantulas also have silk glands in their feet.

Even species that do not build webs to catch prey use silk in several ways: as wrappers for sperm and for fertilized eggs; as a “safety rope”; for nest-building; and as “parachutes” by the young of some species.

Reproduction and Life Cycle

The tiny male of the Golden orb weaver (*Nephila clavipes*) (near the top of the leaf) is protected from the female by his producing the right vibrations in the web, and may be too small to be worth eating.

Spiders reproduce sexually and fertilization is internal but indirect, in other words the sperm is not inserted into the female's body by the male's genitals but by an intermediate stage. Unlike many land-living arthropods, male spiders do not produce ready-made spermatophores (packages of sperm), but spin small sperm webs on to which they ejaculate and then transfer the sperm to special syringe-like structures, palpal bulbs or palpal organs, borne on the tips of the pedipalps of mature males. When a male detects signs of a female nearby he checks whether she is of the same species and whether she is ready to mate; for example in species that produce webs or "safety ropes", the male can identify the species and sex of these objects by "smell".

Spiders generally use elaborate courtship rituals to prevent the large females from eating the small males before fertilization, except where the male is so much smaller that he is not worth eating. In web-weaving species, precise patterns of vibrations in the web are a major part of the rituals, while patterns of touches on the female's body are important in many spiders that hunt actively, and may "hypnotize" the female. Gestures and dances by the male are important for jumping spiders, which have excellent eyesight. If courtship is successful, the male injects his sperm from the palpal bulbs into the female's genital opening, known as the *epigyne*, on the underside of her abdomen. Female's reproductive tracts vary from simple tubes to systems that include seminal receptacles in which females store sperm and release it when they are ready.

Orange spider egg sac hanging from ceiling.

Males of the genus *Tidarren* amputate one of their palps before maturation and enter adult life with one palp only. The palps are 20% of male's body mass in this species, and detaching one of the two improves mobility. In the Yemeni species *Tidarren argo*, the remaining palp is then torn off by the female. The separated palp remains attached to the female's epigynum for about four hours and apparently continues to function independently. In the meantime, the female feeds on the palpless male. In over 60% of cases, the female of the Australian redback spider kills and eats the male after it inserts its second palp into the female's genital opening; in fact, the males co-operate by trying to impale themselves on the females' fangs. Observation shows that most male redbacks never get an opportunity to mate, and the "lucky" ones increase the likely number of offspring by ensuring that the females are well-fed. However, males of most species survive a few matings, limited mainly by their short life spans. Some even live for a while in their mates' webs.

Gasteracantha mammosa spiderlings next to their eggs capsule.

Wolf spider carrying its young on its abdomen.

Females lay up to 3,000 eggs in one or more silk egg sacs, which maintain a fairly constant humidity level. In some species, the females die afterwards, but females of other species protect the sacs by attaching them to their webs, hiding them in nests, carrying them in the chelicerae or attaching them to the spinnerets and dragging them along.

Baby spiders pass all their larval stages inside the egg and hatch as spiderlings, very small and sexually immature but similar in shape to adults. Some spiders care for their young, for example a wolf spider's brood cling to rough bristles on the mother's back, and females of some species respond to the "begging" behaviour of their young by giving them their prey, provided it is no longer struggling, or even regurgitate food.

Like other arthropods, spiders have to molt to grow as their cuticle ("skin") cannot stretch. In some species males mate with newly molted females, which are too weak to be dangerous to the males. Most spiders live for only one to two years, although some tarantulas can live in captivity for over 20 years.

Size

Goliath birdeater (*Theraphosa blondi*), the largest spider.

Spiders occur in a large range of sizes. The smallest, *Patu digua* from Colombia, are less than 0.37 mm (0.015 in) in body length. The largest and heaviest spiders occur among tarantulas, which can have body lengths up to 90 mm (3.5 in) and leg spans up to 250 mm (9.8 in).

Coloration

Only three classes of pigment (ommochromes, bilins and guanine) have been identified in spiders, although other pigments have been detected but not yet characterized. Melanins, carotenoids and pterins, very common in other animals, are apparently absent. In some species, the exocuticle of the legs and prosoma is modified by a tanning process, resulting in brown coloration. Bilins are found, for example, in *Micrommata virescens*, resulting in its green color. Guanine is responsible for the white markings of the European garden spider *Araneus diadematus*. It is in many species accumulated in specialized cells called guanocytes. In genera such as *Tetragnatha*, *Leucauge*, *Argyrodes* or *Theridiosoma*, guanine creates their silvery appearance. While guanine is originally an end-product of protein metabolism, its excretion can be blocked in spiders, leading to an increase in its storage. Structural colors occur in some species, which are the result of the diffraction, scattering or interference of light, for example by modified setae or scales. The white prosoma of *Argiope* results from hairs reflecting the light, *Lycosa* and *Josa* both have areas of modified cuticle that act as light reflectors.

ECOLOGY AND BEHAVIOR

Non-predatory Feeding

A jumping spider seen in Chennai.

Although spiders are generally regarded as predatory, the jumping spider *Bagheera kiplingi* gets over 90% of its food from fairly solid plant material produced by acacias as part of a mutually beneficial relationship with a species of ant.

Juveniles of some spiders in the families Anyphaenidae, Corinnidae, Clubionidae, Thomisidae and Salticidae feed on plant nectar. Laboratory studies show that they do so deliberately and over extended periods, and periodically clean themselves while feeding. These spiders also prefer sugar solutions to plain water, which indicates that they are seeking nutrients. Since many spiders are nocturnal, the extent of nectar consumption by spiders may have been underestimated. Nectar contains amino acids, lipids, vitamins and minerals in addition to sugars, and studies have shown that other spider species live longer when nectar is available. Feeding on nectar avoids the risks of struggles with prey, and the costs of producing venom and digestive enzymes.

Various species are known to feed on dead arthropods (scavenging), web silk, and their own shed exoskeletons. Pollen caught in webs may also be eaten, and studies have shown that young spiders have a better chance of survival if they have the opportunity to eat pollen. In captivity, several spider species are also known to feed on bananas, marmalade, milk, egg yolk and sausages.

Methods of Capturing Prey

The *Phonognatha graeffei* or leaf-curling spider's web serves both as a trap and as a way of making its home in a leaf.

The best-known method of prey capture is by means of sticky webs. Varying placement of webs allows different species of spider to trap different insects in the same area, for example flat horizontal webs trap insects that fly up from vegetation underneath while flat vertical webs trap insects in horizontal flight. Web-building spiders have poor vision, but are extremely sensitive to vibrations.

Females of the water spider *Argyroneta aquatica* build underwater "diving bell" webs that they fill with air and use for digesting prey, molting, mating and raising offspring. They live almost entirely within the bells, darting out to catch prey animals that touch the bell or the threads that anchor it. A few spiders use the surfaces of lakes and ponds as "webs", detecting trapped insects by the vibrations that these cause while struggling.

Net-casting spiders weave only small webs, but then manipulate them to trap prey. Those of the genus *Hyptiotes* and the family Theridiosomatidae stretch their webs and then release them when prey strike them, but do not actively move their webs. Those of the family Deinopidae weave even smaller webs, hold them outstretched between their first two pairs of legs, and lunge and push the webs as much as twice their own body length to trap prey, and this move may increase the webs' area by a factor of up to ten. Experiments have shown that *Deinopis spinosus* has two different techniques for trapping prey: backwards strikes to catch flying insects, whose vibrations it detects; and forward strikes to catch ground-walking prey that it sees. These two techniques have also been observed in other deinopids. Walking insects form most of the prey of most deinopids, but one population of *Deinopis subrufa* appears to live mainly on tipulid flies that they catch with the backwards strike.

Mature female bolas spiders of the genus *Mastophora* build "webs" that consist of only a single "trapeze line", which they patrol. They also construct a bolas made of a single thread, tipped with a large ball of very wet sticky silk. They emit chemicals that resemble the pheromones of moths, and then swing the bolas at the moths. Although they miss on about 50% of strikes, they catch about the same weight of insects per night as web-weaving spiders of similar size. The spiders eat the bolas if they have not made a kill in about 30 minutes, rest for a while, and then make new bolas. Juveniles and adult males are much smaller and do not make bolas. Instead they release different pheromones that attract moth flies, and catch them with their front pairs of legs.

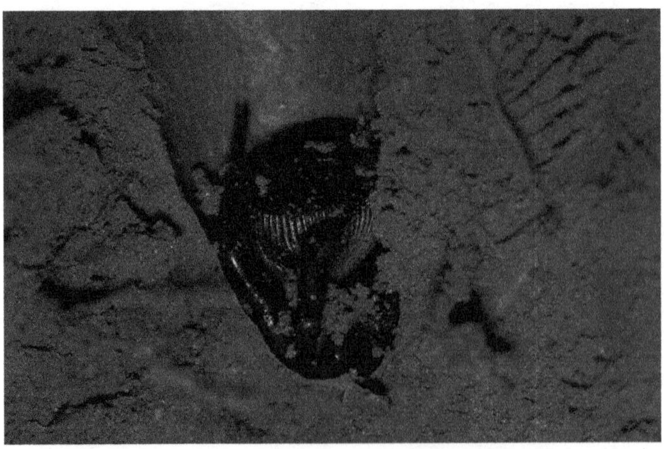

A trapdoor spider in the genus *Cyclocosmia*, an ambush predator.

The primitive Liphistiidae, the "trapdoor spiders" of the family Ctenizidae and many tarantulas are ambush predators that lurk in burrows, often closed by trapdoors and often surrounded by networks of silk threads that alert these spiders to the presence of prey. Other ambush predators do without such aids, including many crab spiders, and a few species that prey on bees, which see ultraviolet, can adjust their ultraviolet reflectance to match the flowers in which they are lurking. Wolf spiders, jumping spiders, fishing spiders and some crab spiders capture prey by chasing it, and rely mainly on vision to locate prey.

Portia uses both webs and cunning, versatile tactics to overcome prey.

Some jumping spiders of the genus *Portia* hunt other spiders in ways that seem intelligent, outflanking their victims or luring them from their webs. Laboratory studies show that *Portia*'s instinctive tactics are only starting points for a trial-and-error approach from which these spiders learn very quickly how to overcome new prey species. However, they seem to be relatively slow "thinkers", which is not surprising, as their brains are vastly smaller than those of mammalian predators.

An ant-mimicking jumping spider.

Ant-mimicking spiders face several challenges: they generally develop slimmer abdomens and false "waists" in the cephalothorax to mimic the three distinct regions (tagmata) of an ant's body; they wave the first pair of legs in front of their heads to mimic antennae, which spiders lack, and to conceal the fact that they have eight legs rather than six; they develop large color patches round one pair of eyes to disguise the fact that they generally have eight simple eyes, while ants have two compound eyes; they cover their bodies with reflective hairs to resemble the shiny bodies of ants. In some spider species, males and females mimic different ant species, as female spiders are usually much larger than males. Ant-mimicking spiders also modify their behavior to resemble that of the target species of ant; for example, many adopt a zig-zag pattern of movement, ant-mimicking jumping spiders avoid jumping, and spiders of the genus *Synemosyna* walk on the outer edges of leaves in the same way as *Pseudomyrmex*. Ant-mimicry in many spiders and other arthropods may be for protection from predators that hunt by sight, including birds, lizards and spiders. However, several ant-mimicking spiders prey either on ants or on the ants' "livestock", such as aphids. When at rest, the ant-mimicking crab spider *Amyciaea* does not closely resemble *Oecophylla*, but while hunting it imitates the behavior of a dying ant to attract worker ants. After a kill, some ant-mimicking spiders hold their victims between themselves and large groups of ants to avoid being attacked.

Defense

Threat display by a Sydney funnel-web spider (*Atrax robustus*).

There is strong evidence that spiders' coloration is camouflage that helps them to evade their major predators, birds and parasitic wasps, both of which have good color vision. Many spider species are colored so as to merge with their most common backgrounds, and some have disruptive coloration, stripes and blotches that break up their outlines. In a few species, such as the Hawaiian happy-face spider, *Theridion grallator*, several coloration schemes are present in a ratio that appears to remain constant, and this may make it more difficult for predators to recognize the species. Most spiders are insufficiently dangerous or unpleasant-tasting for warning coloration to offer much benefit. However, a few species with powerful venoms, large jaws or irritant hairs have patches of warning colors, and some actively display these colors when threatened.

Many of the family Theraphosidae, which includes tarantulas and baboon spiders, have urticating hairs on their abdomens and use their legs to flick them at attackers. These hairs are fine setae (bristles) with fragile bases and a row of barbs on the tip. The barbs cause intense irritation but there is no evidence that they carry any kind of venom. A few defend themselves against wasps by including networks of very robust threads in their webs, giving the spider time to flee while the wasps are struggling with the obstacles. The golden wheeling spider, *Carparachne aureoflava*, of the Namibian desert escapes parasitic wasps by flipping onto its side and cartwheeling down sand dunes.

Social Spiders

A few spider species that build webs live together in large colonies and show social behavior, although not as complex as in social insects. *Anelosimus eximius* (in the family Theridiidae) can form colonies of up to 50,000 individuals. The genus *Anelosimus* has a strong tendency towards sociality: all known American species are social, and species in Madagascar are at least somewhat social. Members of other species in the same family but several different genera have independently developed social behavior. For example, although *Theridion nigroannulatum* belongs to a genus with no other social species, *T. nigroannulatum* build colonies that may contain several thousand individuals that co-operate in prey capture and share food. Other communal spiders include several *Philoponella* species (family Uloboridae), *Agelena consociata* (family Agelenidae) and *Mallos gregalis* (family Dictynidae). Social predatory spiders need to defend their prey against kleptoparasites ("thieves"), and larger colonies are more successful in this. The herbivorous spider *Bagheera kiplingi* lives in small colonies which help to protect eggs and spiderlings. Even widow spiders (genus *Latrodectus*), which are notoriously cannibalistic, have formed small colonies in captivity, sharing webs and feeding together.

Web Types

There is no consistent relationship between the classification of spiders and the types of web they build: species in the same genus may build very similar or significantly different webs. Nor is there much correspondence between spiders' classification and the chemical composition of their silks. Convergent evolution in web construction, in other words use of similar techniques by remotely related species, is rampant. Orb web designs and the spinning behaviors that produce them are the best understood. The basic radial-then-spiral sequence visible in orb webs and the sense of direction required to build them may have been inherited from the common ancestors of most spider groups. However, the majority of spiders build non-orb webs. It used to be thought that the sticky orb web was an evolutionary innovation resulting in the diversification of the Orbiculariae.

Now, however, it appears that non-orb spiders are a sub-group that evolved from orb-web spiders, and non-orb spiders have over 40% more species and are four times as abundant as orb-web spiders. Their greater success may be because sphecid wasps, which are often the dominant predators of spiders, much prefer to attack spiders that have flat webs.

The large orb web of *Araneus diadematus* (European garden spider).

Orb Webs

Nephila clavata, a golden orb weaver.

About half the potential prey that hit orb webs escape. A web has to perform three functions: intercepting the prey (intersection), absorbing its momentum without breaking (stopping), and trapping the prey by entangling it or sticking to it (retention). No single design is best for all prey. For example: wider spacing of lines will increase the web's area and hence its ability to intercept prey, but reduce its stopping power and retention; closer spacing, larger sticky droplets and thicker lines would improve retention, but would make it easier for potential prey to see and avoid the web, at least during the day. However, there are no consistent differences between orb webs built for use during the day and those built for use at night. In fact, there is no simple relationship between orb web design features and the prey they capture, as each orb-weaving species takes a wide range of prey.

The hubs of orb webs, where the spiders lurk, are usually above the center, as the spiders can move downwards faster than upwards. If there is an obvious direction in which the spider can retreat to avoid its own predators, the hub is usually offset towards that direction.

Horizontal orb webs are fairly common, despite being less effective at intercepting and retaining prey and more vulnerable to damage by rain and falling debris. Various researchers have suggested that horizontal webs offer compensating advantages, such as reduced vulnerability to wind damage; reduced visibility to prey flying upwards, because of the back-lighting from the sky; enabling oscillations to catch insects in slow horizontal flight. However, there is no single explanation for the common use of horizontal orb webs.

Spiders often attach highly visible silk bands, called decorations or stabilimenta, to their webs. Field research suggests that webs with more decorative bands captured more prey per hour. However, a laboratory study showed that spiders reduce the building of these decorations if they sense the presence of predators.

There are several unusual variants of orb web, many of them convergently evolved, including: attachment of lines to the surface of water, possibly to trap insects in or on the surface; webs with twigs through their centers, possibly to hide the spiders from predators; "ladder-like" webs that appear most effective in catching moths. However, the significance of many variations is unclear.

In 1973, Skylab 3 took two orb-web spiders into space to test their web-spinning capabilities in zero gravity. At first, both produced rather sloppy webs, but they adapted quickly.

Tangleweb Spiders (cobweb spiders)

Members of the family Theridiidae weave irregular, tangled, three-dimensional webs, popularly known as cobwebs. There seems to be an evolutionary trend towards a reduction in the amount of sticky silk used, leading to its total absence in some species. The construction of cobwebs is less stereotyped than that of orb-webs, and may take several days.

A funnel web.

Other types of Webs

The Linyphiidae generally make horizontal but uneven sheets, with tangles of stopping threads above. Insects that hit the stopping threads fall onto the sheet or are shaken onto it by the spider, and are held by sticky threads on the sheet until the spider can attack from below.

Evolution

Fossil Record

Spider preserved in amber.

Although the fossil record of spiders is considered poor, almost 1000 species have been described from fossils.Because spiders' bodies are quite soft, the vast majority of fossil spiders have been found preserved in amber. The oldest known amber that contains fossil arthropods dates from 130 million years ago in the Early Cretaceous period. In addition to preserving spiders' anatomy in very fine detail, pieces of amber show spiders mating, killing prey, producing silk and possibly caring for their young. In a few cases, amber has preserved spiders' egg sacs and webs, occasionally with prey attached; the oldest fossil web found so far is 100 million years old. Earlier spider fossils come from a few lagerstätten, places where conditions were exceptionally suited to preserving fairly soft tissues.

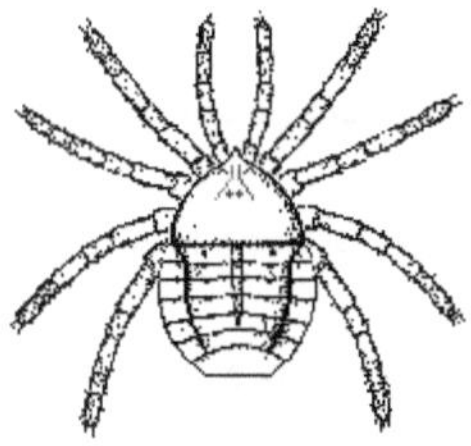

Palaeotarbus jerami, a trigonotarbid and the oldest known exclusively terrestrial arachnid.

The oldest known exclusively terrestrial arachnid is the trigonotarbid *Palaeotarbus jerami*, from about 420 million years ago in the Silurian period, and had a triangular cephalothorax and segmented abdomen, as well as eight legs and a pair of pedipalps. *Attercopus fimbriunguis*, from 386 million years ago in the Devonian period, bears the earliest known silk-producing spigots, and was therefore hailed as a spider at the time of its discovery.However, these spigots may have been mounted on the underside of the abdomen rather than on spinnerets, which are modified appendages and whose mobility is important in the building of webs. Hence *Attercopus* and the similar Permian arachnid *Permarachne* may not have been true spiders, and probably used silk for lining nests or producing egg-cases rather than for building webs. The largest known fossil spider as of 2011 is the araneid *Nephila jurassica*, from about 165 million years ago, recorded from Daohuogo, Inner Mongolia in China. Its body length is almost 25 mm, (i.e., almost one inch).

Several Carboniferous spiders were members of the Mesothelae, a primitive group now represented only by the Liphistiidae. The mesothelid *Palęothele montceauensis*, from the Late Carboniferous over 299 million years ago, had five spinnerets. Although the Permian period 299 to 251 million years ago saw rapid diversification of flying insects, there are very few fossil spiders from this period.

The main groups of modern spiders, Mygalomorphae and Araneomorphae, first appear in the Triassic well before 200 million years ago. Some Triassic mygalomorphs appear to be members of the family Hexathelidae, whose modern members include the notorious Sydney funnel-web spider, and their spinnerets appear adapted for building funnel-shaped webs to catch jumping insects. Araneomorphae account for the great majority of modern spiders, including those that weave the familiar orb-shaped webs.

The Jurassic and Cretaceous periods provide a large number of fossil spiders, including representatives of many modern families.

Family Tree

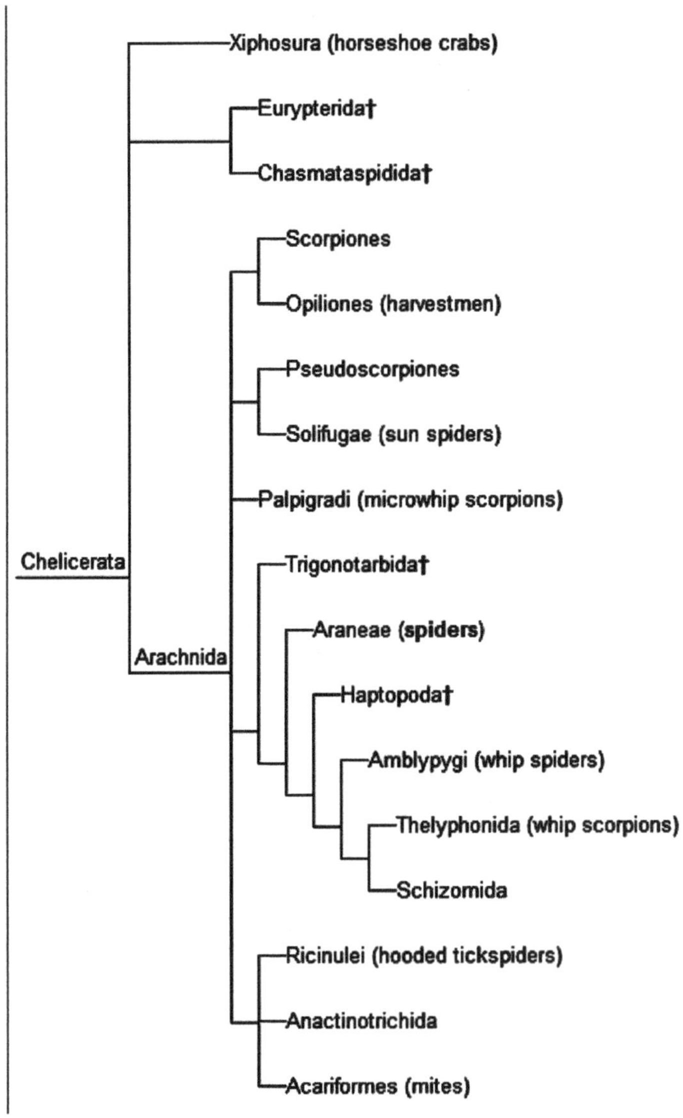

Shultz (2007)'s evolutionary family tree of arachnids – † marks extinct groups.

It is now agreed that spiders (Araneae) are monophyletic (i.e., members of a group of organisms that form a clade, consisting of a last common ancestor and all of its descendants). There has been debate about what their closest evolutionary relatives are, and how all of these evolved from the ancestral chelicerates, which were marine animals. The cladogram on the right is based on J. W. Shultz' analysis (2007). Other views include proposals that: scorpions are more closely related to the extinct marine scorpion-like eurypterids than to spiders; spiders and Amblypygi are a monophyletic group. The appearance of several multi-way branchings in the tree on the right shows that there are still uncertainties about relationships between the groups involved.

Arachnids lack some features of other chelicerates, including backward-pointing mouths and gnathobases ("jaw bases") at the bases of their legs; both of these features are part of the ancestral arthropod feeding system. Instead, they have mouths that point forwards and downwards, and all have some means of breathing air.Spiders (Araneae) are distinguished from other arachnid groups by several characteristics, including spinnerets and, in males, pedipalps that are specially adapted for sperm transfer.

Taxonomy

Spiders are divided into two suborders, Mesothelae and Opisthothelae, of which the latter contains two infraorders, Mygalomorphae and Araneomorphae. Nearly 46,000 living species of spiders (order Araneae) have been identified and are currently grouped into about 114 families and about 4,000 genera by arachnologists.

<table>
<tr><th></th><th colspan="3">Spider diversity (numbers are approximate)</th><th colspan="3">Features</th><th></th></tr>
<tr><th>Suborder/ Infraorder</th><th>Species</th><th>Genera</th><th>Families</th><th>Segmented plates on top of abdomen</th><th>Ganglia in abdomen</th><th>Spinnerets</th><th>Striking direction of fangs</th></tr>
<tr><td>Mesothelae</td><td>87</td><td>5</td><td>1</td><td>Yes</td><td>Yes</td><td>Four pairs, in some species one pair fused, under middle of abdomen</td><td rowspan="2">Downwards and forwards</td></tr>
<tr><td>Opisthothelae: Mygalomorphae</td><td>2,600</td><td>300</td><td>15</td><td rowspan="2">Only in some fossils</td><td rowspan="2">No</td><td rowspan="2">One, two or three pairs under rear of abdomen</td></tr>
<tr><td>Opisthothelae: Araneomorphae</td><td>37,000</td><td>3,400</td><td>93</td><td>From sides to center, like pincers</td></tr>
</table>

Mesothelae

The only living members of the primitive Mesothelae are the family Liphistiidae, found only in Southeast Asia, China, and Japan. Most of the Liphistiidae construct silk-lined burrows with thin trapdoors, although some species of the genus *Liphistius* build camouflaged silk tubes with a second trapdoor as an emergency exit. Members of the genus *Liphistius* run silk "tripwires" outwards from their tunnels to help them detect approaching prey, while those of genus *Heptathela* do not and instead rely on their built-in vibration sensors. Spiders of the genus *Heptathela* have no venom glands although they do have venom gland outlets on the fang tip.

Ryuthela secundaria, a member of the Liphistiidae

The extinct families Arthrolycosidae, found in Carboniferous and Permian rocks, and Arthromygalidae, so far found only in Carboniferous rocks, have been classified as members of the Mesothelae.

Mygalomorphae

A Mexican red-kneed tarantula *Brachypelma smithi*.

The Mygalomorphae, which first appeared in the Triassic period, are generally heavily built and hairy, with large, robust chelicerae and fangs. Well-known examples include tarantulas, ctenizid trapdoor spiders and the Australasian funnel-web spiders. Most spend the majority of their time in burrows, and some run silk tripwires out from these, but a few build webs to capture prey. However, mygalomorphs cannot produce the pirifom silk that the Araneomorphae use as instant adhesive to glue silk to surfaces or to other strands of silk, and this makes web construction more difficult for mygalomorphs. Since mygalomorphs rarely "balloon" by using air currents for transport, their populations often form clumps. In addition to arthropods, mygalomorphs are capable of preying on frogs, small mammals, lizards, and snails.

Araneomorphae

Leucauge venusta, an orb-web spider.

In addition to accounting for over 90% of spider species, the Araneomorphae, also known as the "true spiders", include orb-web spiders, the cursorial wolf spiders, and jumping spiders, as well as the only known herbivorous spider, *Bagheera kiplingi*. They are distinguished by having fangs that oppose each other and cross in a pinching action, in contrast to the Mygalomorphae, which have fangs that are nearly parallel in alignment.

Spiders and people.

Spider Bites

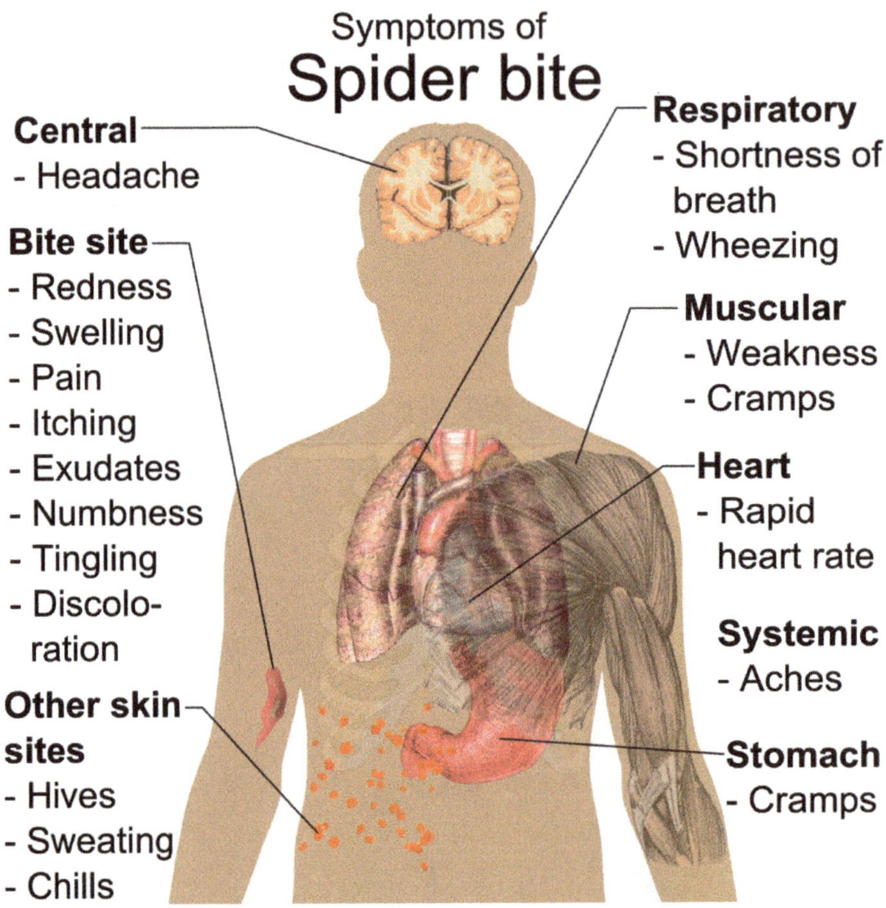

All symptoms associated with toxic spider bites

Although spiders are widely feared, only a few species are dangerous to people. Spiders will only bite humans in self-defense, and few produce worse effects than a mosquito bite or bee-sting. Most of those with medically serious bites, such as recluse spiders and widow spiders, would rather flee and bite only when trapped, although this can easily arise by accident. Funnel web spiders' defensive tactics include fang display and their venom, although they rarely inject much, has resulted in 13 known human deaths over 50 years. They have been deemed to be the world's most dangerous spiders on clinical and venom toxicity grounds, though this claim has also been attributed to the Brazilian wandering spider, due to much more frequent accidents.

There were about 100 reliably reported deaths from spider bites in the 20th century, compared to about 1,500 from jellyfish stings. Many alleged cases of spider bites may represent incorrect diagnoses,which would make it more difficult to check the effectiveness of treatments for genuine bites.

Benefits to Humans

Cooked tarantula spiders are considered a delicacy in Cambodia.

Spider venoms may be a less polluting alternative to conventional pesticides, as they are deadly to insects but the great majority are harmless to vertebrates. Australian funnel web spiders are a promising source, as most of the world's insect pests have had no opportunity to develop any immunity to their venom, and funnel web spiders thrive in captivity and are easy to "milk". It may be possible to target specific pests by engineering genes for the production of spider toxins into viruses that infect species such as cotton bollworms.

The Ch'ol Maya use a beverage created from the tarantula species *Brachypelma vagans* for the treatment of a condition they term 'tarantula wind', the symptoms of which include chest pain, asthma and coughing.

Possible medical uses for spider venoms are being investigated, for the treatment of cardiac arrhythmia,Alzheimer's disease, strokes, and erectile dysfunction. The peptide GsMtx-4, found in the venom of Brachypelma vagans, is being researched to determine whether or not it could effectively be used for the treatment of cardiac arrhythmia, muscular dystrophy or glioma.Because spider silk is both light and very strong, attempts are being made to produce it in goats' milk and in the leaves of plants, by means of genetic engineering.

Spiders can also be used as food. Cooked tarantula spiders are considered a delicacy in Cambodia, and by the Piaroa Indians of southern Venezuela – provided the highly irritant hairs, the spiders' main defense system, are removed first.

Arachnophobia

Arachnophobia is a specific phobia—it is the abnormal fear of spiders or anything reminiscent of spiders, such as webs or spider-like shapes. It is one of the most common specific phobias, and some statistics show that 50% of women and 10% of men show symptoms. It may be an exaggerated form of an instinctive response that helped early humans to survive, or a cultural phenomenon that is most common in predominantly European societies.

Spiders in Symbolism and Culture

This Moche ceramic depicts a spider, and dates from around 300 CE.

Spiders have been the focus of stories and mythologies of various cultures for centuries. They have symbolized patience due to their hunting technique of setting webs and waiting for prey, as well as mischief and malice due to their venomous bites. The Italian tarantella is a dance to rid the young woman of the lustful effects of a spider bite.

Web-spinning also caused the association of the spider with creation myths, as they seem to have the ability to produce their own worlds. Dreamcatchers are depictions of spiderwebs. The Moche people of ancient Peru worshipped nature. They placed emphasis on animals and often depicted spiders in their art.

SCORPION

Scorpions are predatory arachnids of the order **Scorpiones**. They have eight legs and are easily recognized by the pair of grasping pedipalps and the narrow, segmented tail, often carried in a characteristic forward curve over the back, ending with a venomous stinger. Scorpions range in size from 9 mm / 0.3 in. (*Typhlochactas mitchelli*) to 23 cm / 9 in. (*Heterometrus swammerdami*).

The evolutionary history of scorpions goes back to the Silurian era 430 million years ago. They have adapted to a wide range of environmental conditions and can now be found on all continents except Antarctica. Scorpions number about 1750 described species, with 13 extant families recognised to date. Only about 25 of these species are known to have venom capable of killing a human being.[1] The taxonomy has undergone changes and is likely to change further, as genetic studies are bringing forth new information.

Scorpion stings are painful but are usually harmless to humans. For stings from species found in North America, no treatment is normally needed for healthy adults, although medical care should be sought for children and for the elderly. More harmful stings from species found in South America, Africa, and western Asia may require medical attention.

Etymology

The word *scorpion* is thought to have originated in Middle English between 1175 and 1225 AD from Old French *scorpion*, or from Italian*scorpione*, both derived from the Latin word *scorpius*, which is the romanization of the Greek word σκορπίος – *skorpíos*.

Geographical Distribution

Scorpions are found on all major land masses except Antarctica. Scorpions did not occur naturally in Great Britain, Japan, South Korea, New Zealand and some of the islands in Oceania, but now have been accidentally introduced in some of these places by human trade and commerce. The greatest diversity of scorpions in the Northern Hemisphere is to be found in the subtropical areas lying between latitudes 23° N and 38° N. Above these latitudes, the diversity decreases, with the northernmost natural occurrence of scorpions being the northern scorpion *Paruroctonus boreus* at Medicine Hat, Alberta, Canada 50° N.

Today, scorpions are found in virtually every terrestrial habitat, including high-elevation mountains, caves and intertidal zones, with the exception of boreal ecosystems, such as the tundra, high-altitude taiga and the permanently snow-clad tops of some mountains. As regards microhabitats, scorpions may be ground-dwelling, tree-living, lithophilic (rock-loving) or psammophilic (sand-loving); some species, such as *Vaejovis janssi*, are versatile and found in every type of habitat in Baja California, while others occupy specialized niches such as *Euscorpius carpathicus*, which occupies the littoral zone of the shore.

Five colonies of scorpions (*Euscorpius flavicaudis*) have established themselves in Sheerness on the Isle of Sheppey in the United Kingdom.This small population has been resident since the 1860s, having probably arrived with imported fruit from Africa. This scorpion species is small and completely harmless to humans. At just over 51° N, this marks the northernmost limit where scorpions live in the wild.

Classification

There are thirteen families and about 1,750 described species and subspecies of scorpions. In addition, there are 111 described taxa of extinct scorpions.

This classification is based on that of Soleglad & Fet (2003), which replaced the older, unpublished classification of Stockwell. Additional taxonomic changes are from papers by Soleglad et al. (2005).

Systematics

The following classification covers extant taxa to the rank of family.

Order Scorpiones

- Infraorder Orthosterni Pocock, 1911
- Parvorder Pseudochactida Soleglad et Fet, 2003
- Superfamily Pseudochactoidea Gromov, 1998
- Family Pseudochactidae Gromov, 1998
- Parvorder Buthida Soleglad et Fet, 2003
- Superfamily Buthoidea C. L. Koch, 1837
- Family Buthidae C. L. Koch, 1837 (thick-tailed scorpions)
- Family Microcharmidae Lourenço, 1996
- Parvorder Chaerilida Soleglad et Fet, 2003
- Superfamily Chaeriloidea Pocock, 1893
- Family Chaerilidae Pocock, 1893
- Parvorder Iurida Soleglad et Fet, 2003
- Superfamily Chactoidea Pocock, 1893
- Family Chactidae Pocock, 1893
- Family Euscorpiidae Laurie, 1896
- Family Superstitioniidae Stahnke, 1940
- Family Vaejovidae Thorell, 1876
- Superfamily Iuroidea Thorell, 1876
- Family Caraboctonidae Kraepelin, 1905 (hairy scorpions)
- Family Iuridae Thorell, 1876
- Superfamily Scorpionoidea Latreille, 1802
- Family Bothriuridae Simon, 1880
- Family Hemiscorpiidae Pocock, 1893 (= Ischnuridae, =Liochelidae) (rock scorpions, creeping scorpions, or tree scorpions)
- Family Scorpionidae Latreille, 1802 (burrowing scorpions or pale-legged scorpions)

Fossil Record

Scorpions have been found in many fossil records, including marine Silurian and estuarine Devonian deposits, coal deposits from the Carboniferous Period and in amber. The oldest known scorpions lived around 430 million years ago in the Silurian period. Though once believed to have lived on the bottom of shallow tropical seas, early scorpions are now believed to have been terrestrial and to have washed into marine settings together with plant matter. These first scorpions were believed to have had gills instead of the present forms' book lungs though this has subsequently been refuted. The oldest Gondwanan scorpiones (*Gondwanascorpio*) comprise the earliest known terrestrial animals from Gondwana. Currently, 111 fossil species of scorpion are known.Unusually for arachnids, there are more species of Palaeozoic scorpion than Mesozoic or Cenozoic ones.

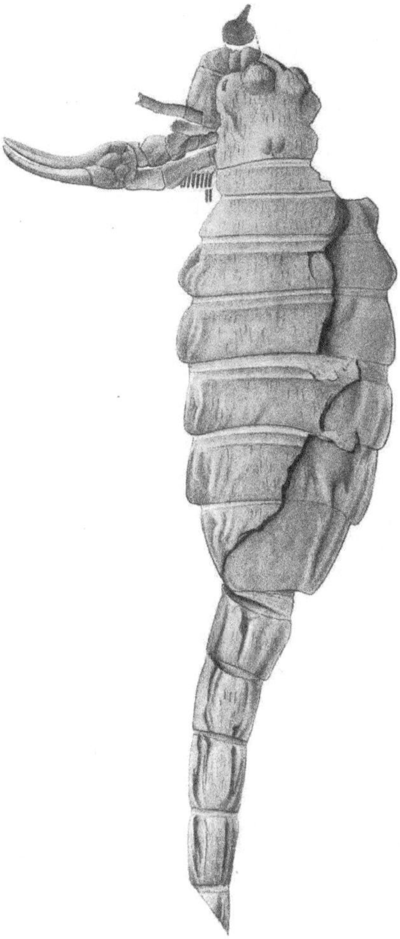

Proscorpius osborni, fossil scorpion initially thought to be a eurypterid.

The eurypterids, marine creatures that lived during the Palaeozoic era, share several physical traits with scorpions and may be closely related to them. Various species of Eurypterida could grow to be anywhere from 10 centimetres (3.9 in) to 2.5 metres (8.2 ft) in length. However, they exhibit anatomical differences marking them off as a group distinct from their Carboniferous and Recent relatives. Despite this, they are commonly referred to as "sea scorpions". Their legs are thought to have been short, thick, tapering and to have ended in a single strong claw; it appears that they were well-adapted for maintaining a secure hold upon rocks or seaweed against the wash of waves, like the legs of a shore crab. Cladistic analyses have supported the idea that the eurypterids are a distinct group from the scorpions.

Morphology

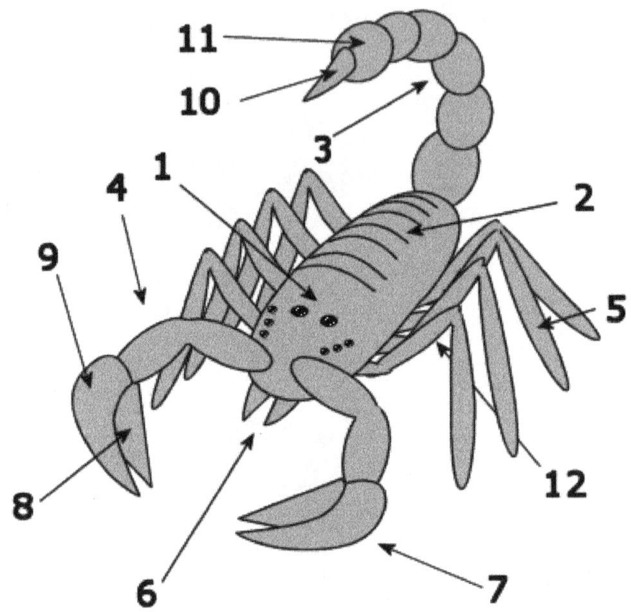

Scorpion anatomy:

1 = Cephalothorax or *Prosoma*;

2 = Abdomen or *Mesosoma*;

3 = Tail or *Metasoma*;

4 = Claws or *Pedipalps*

5 = Legs;

6 = Mouth parts or *Chelicerae*;

7 = pincers or Chelae;

8 = Moveable claw or *Tarsus*;

9 = Fixed claw or *Manus*;

10 = Stinger or *Aculeus*;

11 = *Telson* (follows anus in previous joint).

Front of scorpion.

Stinger of an Arizona bark scorpion.

Ventral view of an unidentified scorpion species where the pectines can easily be observed with a comb like structure in an inverted V shape.

The body of a scorpion is divided into two parts (tagmata): the head (cephalothorax) and the abdomen (opisthosoma), which is subdivided into a broad anterior (mesosoma), or preabdomen, and a narrow taillike posterior (metasoma), or postabdomen.

Cephalothorax

The cephalothorax, also called the *prosoma*, comprises the carapace, eyes, chelicerae (mouth parts), pedipalps (the pedipalps of scorpions have chelae, commonly called claws or pincers) and four pairs of walking legs. The scorpion's exoskeleton is thick and durable, providing good protection from predators. Scorpions have two eyes on the top of the cephalothorax, and usually two to five pairs of eyes along the front corners of the cephalothorax. The position of the eyes on the cephalothorax depends in part on the hardness or softness of the soil upon which they spend their lives.

The pedipalp is a segmented, chelate (clawed) appendage used for prey immobilization, defense and sensory purposes. The segments of the pedipalp (from closest to the body outwards) are coxa, trochanter, femur (humerus), patella, tibia (including the fixed claw and the manus) and tarsus (moveable claw). A scorpion has darkened or granular raised linear ridges, called "keels" or *carinae* on the pedipalp segments and on other parts of the body, which are useful taxonomically.

Mesosoma

The abdomen, also called the *opisthosoma*, consists of seven segments (somites), each covered dorsally by a sclerotosed plate (tergum) and also ventrally for segments 3 to 7. The first abdominal segment bears a pair of genital opercula covering the gonopore. Segment 2 consists of the basal plate with the pectines, which are a pair of limbs transformed into sensory organs. Segments 3 to 7 each have a pair of spiracles, the openings for the scorpion's respiratory organs, known as book lungs. The spiracle openings may be slits, circular, elliptical, or oval.

Metasoma

The metasoma, the scorpion's tail, comprises five caudal segments (the first tail segment looks like a last mesosoman segment) and the sixth bearing the telson (the sting). The telson, in turn, consists of the vesicle, which holds a pair of venom glands, and the hypodermic aculeus, the venom-injecting barb.

On rare occasions, scorpions can be born with two metasomata (tails). Two-tailed scorpions are not a different species, merely a genetic abnormality.

Fluorescence

Scorpions are also known to glow a vibrant blue-green when exposed to certain wavelengths of ultraviolet light such as that produced by a black light, due to the presence of fluorescent chemicals in the cuticle. One fluorescent component is now known to be beta-carboline. A hand-held UV lamp has long been a standard tool for nocturnal field surveys of these animals. Fluorescence occurs as a result of sclerotisation and increases in intensity with each successive instar. This fluorescence may have an active role in scorpion light detection.

This black scorpion appears light-blue under a black light.

Biology

Scorpions prefer areas where the temperatures range from 20 to 37 °C (68 to 99 °F), but may survive temperatures ranging from well below freezing to desert heat. Scorpions of the genus *Scorpiops* living in high Asian mountains, bothriurid scorpions from Patagonia and small *Euscorpius* scorpions from Central Europe can all survive winter temperatures of about −25 °C (−13 °F). In Repetek (Turkmenistan), seven species of scorpion (of which *Pectinibuthus birulai* is endemic) live in temperatures varying from −31 to 50 °C (−24 to 122 °F).

They are nocturnal and fossorial, finding shelter during the day in the relative cool of underground holes or undersides of rocks, and emerging at night to hunt and feed. Scorpions exhibit photophobic behavior, primarily to evade detection by predators such as birds, centipedes, lizards, mice, opossums, and rats.

Scorpions are opportunistic predators of small arthropods, although the larger kinds have been known to kill small lizards and mice. The large pincers are studded with highly sensitive tactile hairs, and the moment an insect touches these, they use their chelae (pincers) to catch the prey. Depending on the toxicity of their venom and size of their claws, they will then either crush the prey or inject it with neurotoxic venom. This will kill or paralyze the prey so the scorpion can eat it. Scorpions have an unusual style of eating using chelicerae, small claw-like structures that protrude from the mouth that are unique to the Chelicerata among arthropods. The chelicerae, which are very sharp, are used to pull small amounts of food off the prey item for digestion into a *pre-oral cavity* below the chelicerae and carapace. Scorpions can ingest food only in a liquid form; they have external digestion. The digestive juices from the gut are egested onto the food and the digested food sucked in liquid form. Any solid indigestible matter (fur, exoskeleton, etc.) is trapped by setae in the pre-oral cavity, which is ejected by the scorpion.

Scorpions can consume huge amounts of food at one sitting. They have a very efficient food storage organ and a very low metabolic rate combined with a relatively inactive lifestyle. This enables scorpions to survive long periods when deprived of food; some are able to survive 6 to 12 months of starvation. Scorpions excrete very little; their waste consists mostly of insoluble nitrogenous compounds, such as xanthine, guanine and uric acid.

Reproduction

Most scorpions reproduce sexually, and most species have male and female individuals. However, some species, such as *Hottentotta hottentotta*, *Hottentotta caboverdensis*, *Liocheles australasiae*, *Tityus columbianus*, *Tityus metuendus*, *Tityus serrulatus*, *Tityus stigmurus*, *Tityus trivittatus* and *Tityus urugayensis*, reproduce through parthenogenesis, a process in which unfertilised eggs develop into living embryos. Parthenogenic reproduction starts following the scorpion's final moult to maturity and continues thereafter.

Sexual reproduction is accomplished by the transfer of a spermatophore from the male to the female; scorpions possess a complex courtship and mating ritual to effect this transfer. Mating starts with the male and female locating and identifying each other using a mixture of pheromones and vibrational communication. Once they have satisfied the other that they are of opposite sex and of the correct species, mating can commence.

The courtship starts with the male grasping the female's pedipalps with his own; the pair then perform a "dance" called the "*promenade à deux*". In this "dance", the male leads the female around searching for a suitable place to deposit his spermatophore. The courtship ritual can involve several other behaviours, such as juddering and a cheliceral kiss, in which the male's chelicerae – pincers – grasp the female's in a smaller more intimate version of the male's grasping the female's pedipalps and in some cases injecting a small amount of his venom into her pedipalp or on the edge of her cephalothorax, probably as a means of pacifying the female.

When the male has identified a suitable location, he deposits the spermatophore and then guides the female over it. This allows the spermatophore to enter her genital opercula, which triggers release of the sperm, thus fertilising the female. The mating process can take from 1 to 25+ hours and depends on the ability of the male to find a suitable place to deposit his spermatophore. If mating continues too long, the female may lose interest, ending the process.

Once the mating is complete, the male and female will separate. The male will generally retreat quickly, most likely to avoid being cannibalized by the female, although sexual cannibalism is infrequent with scorpions.

Pregnant scorpion.

Birth and Development

Compsobuthus werneri female with young.

Unlike the majority of species in the class Arachnida, which are oviparous, scorpions seem to be universally ovoviviparous. The young are born one by one after hatching and expelling the embryonic membrane, if any, and the brood is carried about on its mother's back until the young have undergone at least one moult. Before the first moult, scorplings cannot survive naturally without the mother, since they depend on her for protection and to regulate their moisture levels. Especially in species that display more advanced sociability (e.g. *Pandinus* spp.), the young/mother association can continue for an extended period of time. The size of the litter depends on the species and environmental factors, and can range from two to over a hundred scorplings. The average litter however, consists of around 8 scorplings.

The young generally resemble their parents. Growth is accomplished by periodic shedding of the exoskeleton (ecdysis). A scorpion's developmental progress is measured in instars (how many moults it has undergone). Scorpions typically require between five and seven moults to reach maturity. Moulting commences with a split in the old exoskeleton just below the edge of the carapace (at the front of the prosoma). The scorpion then emerges from this split; the pedipalps and legs are first removed from the old exoskeleton, followed eventually by the metasoma. When it emerges, the scorpion's new exoskeleton is soft, making the scorpion highly vulnerable to attack. The scorpion must constantly stretch while the new exoskeleton hardens to ensure that it can move when the hardening is complete. The process of hardening is called sclerotisation. The new exoskeleton does not fluoresce; as sclerotisation occurs, the fluorescence gradually returns.

Relationship with Humans

Sting and Venom

All known scorpion species possess venom and use it primarily to kill or paralyze their prey so that it can be eaten. In general, it is fast-acting, allowing for effective prey capture. However, as a general rule, they will kill their prey with brute force if they can, as opposed to using venom. It is also used as a defense against predators. The venom is a mixture of compounds (neurotoxins, enzyme inhibitors, etc.) each not only causing a different effect but possibly also targeting a specific animal. Each compound is made and stored in a pair of glandular sacs and is released in a quantity regulated by the scorpion itself. Of the 1,000+ known species of scorpion, only 25 have venom that is deadly to humans; most of those belong to the family Buthidae (including *Leiurus quinquestriatus*, *Hottentotta*, *Centruroides* and *Androctonus*).

Prevention

According to the United States National Institute for Occupational Safety and Health (NIOSH), the following steps should be taken to prevent scorpion stings:

- Wearing long sleeves and pants
- Wearing leather gloves
- Shaking out clothing or shoes before putting them on.
- Workers with a history of severe allergic reactions to insect bites or stings should consider carrying an epinephrine auto injector (EpiPen) and should wear a medical identification bracelet or necklace stating their allergy.

Treatment

First aid for scorpion stings is generally symptomatic. It includes strong analgesia, either systemic (opiates or paracetamol) or locally applied (such as a cold compress). Hypertensive crises are treated with anxiolytics and vasodilators. Scorpion envenomation with high morbidity and mortality is usually due to either excessive autonomic activity and cardiovascular toxic effects or neuromuscular toxic effects. Antivenom is the specific treatment for scorpion envenomation combined with supportive measures including vasodilators in patients with cardiovascular toxic effects and benzodiazepines when there is neuromuscular involvement. Although rare, severe hypersensitivity reactions including anaphylaxis to scorpion antivenin (SAV) are possible.

Medical Use

The deathstalker has powerful venom.

Short-chain scorpion toxins constitute the largest group of potassium (K^+) channel blocking peptides; an important physiological role of the KCNA3 channel, also known as $K_V1.3$, is to help maintain large electrical gradients for the sustained transport of ions such as Ca^{2+} that controls T lymphocyte (T cell) proliferation. Thus $K_V1.3$ blockers could be potential immunosuppressants for the treatment of autoimmune disorders (such as rheumatoid arthritis, inflammatory bowel disease and multiple sclerosis).

The venom of *Uroplectes lineatus* is clinically important in dermatology.

Toxins being investigated include the following:

- Chlorotoxin is a 36–amino acid peptide found in the venom of the deathstalker scorpion (*Leiurus quinquestriatus*) that blocks small-conductance chloride channels. The fact that chlorotoxin binds preferentially to glioma cells has allowed the development of new methods, that still are under investigation, for the treatment and diagnosis of several types of cancer.

- Maurotoxin from the venom of the Tunisian *Scorpio maurus*
- A number of antimicrobial peptides have also been found in the venom of *Mesobuthus eupeus*. Meucin-13 and meucin-18 exhibited extensive cytolytic effects on bacteria, fungi, and yeasts. Furthermore, meucin-24 and meucin-25, first identified from genetic sequences expressed in their venom gland, were shown to selectively kill *Plasmodium falciparum* and inhibit the development of *Plasmodium berghei*, both malaria parasites, but do not harm mammalian cells. These two venom-derived proteins are therefore attractive candidates for the development of anti-malarial drugs.

Production

Scorpions for use in the pharmaceutical industry are collected from the wild in Pakistan. Farmers in the Thatta District are paid about US$100 for each 40 gram scorpion and 60 gram specimens are reported to fetch at least US$50,000. The trade is reported to be illegal but thriving.

Consumption

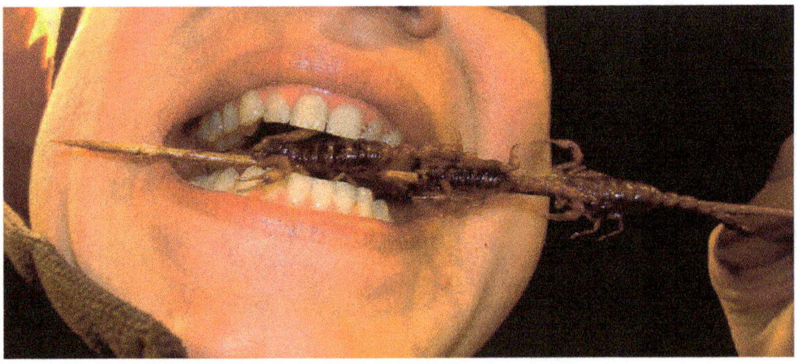

Eating scorpions in Beijing, China.

Fried scorpion is a traditional dish from Shandong, China.

Scorpion and snake wines are used as analgesics.

In Culture

- One of earliest occurrences of the scorpion in culture is its inclusion, as *Scorpio*, in the twelve signs of the series of constellations known as the Zodiac by Babylonian astronomers during the Chaldean period.
- In South Africa and South Asia, the scorpion is a significant animal culturally, appearing as a motif in art, especially in Islamic art in the Middle East. A

scorpion motif is often woven into Turkish kilim flatweave carpets, for protection from their sting. The scorpion is perceived both as an embodiment of evil and a protective force that counters evil, such as a dervish's powers to combat evil. In another context, the scorpion portrays human sexuality. Scorpions are used in folk medicine in South Asia especially in antidotes for scorpion stings.

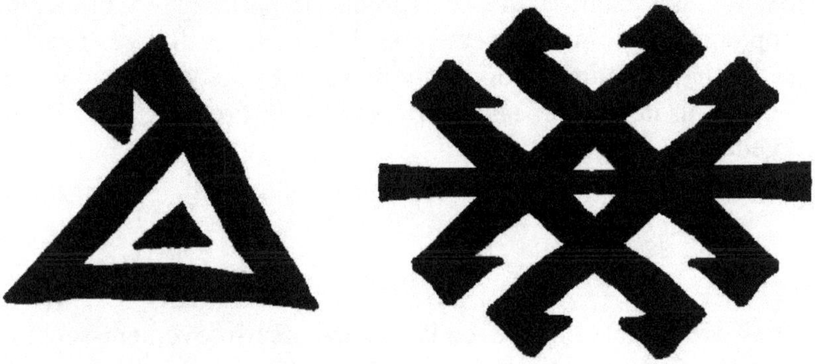

Scorpion motif is often woven into Turkish kilim flatweave carpets, for protection from their sting (2 examples).

- In ancient Egypt the goddess Serket was often depicted as a scorpion, one of several goddesses who protected the Pharaoh.
- Surrealist filmmaker Luis Buñuel makes notable symbolic use of scorpions in his 1930 classic *L'Age d'or* (*The Golden Age*).

REFERENCES

1. Currently valid spider genera and species". World Spider Catalog. Natural History Museum Bern. Retrieved 2016-12-19.
2. Sebastin PA & Peter KV (eds.). (2009) Spiders of India. Universities Press/ Orient Blackswan. ISBN 978-81-7371-641-6
3. Foelix, Rainer F. (1996). Biology of Spiders. 198 Madison Ave. NY, New York, 10016: Oxford University Press. p. 3. ISBN 0-19-509593-6.
4. Meehan, Christopher J.; Olson, Eric J.; Reudink, Matthew W.; Kyser, T. Kurt; Curry, Robert L. (2009). "Herbivory in a spider through exploitation of an ant–plant mutualism". Current Biology. 19 (19): R892–3. doi:10.1016/j.cub.2009.08.049. PMID 19825348.
5. Coddington, J. A. & Levi, H. W. (1991). "Systematics and Evolution of Spiders (Araneae)". Annu. Rev. Ecol. Syst. 22: 565–592. doi:10.1146/annurev.es.22.110191.003025.
6. Barghusen, L. E.; Claussen, D. L.; Anderson, M. S.; Bailer, A. J. (1 February 1997). "The effects of temperature on the web-building behaviour of the common house spider, Achaearanea tepidariorum". Functional Ecology. 11 (1): 4–10. doi:10.1046/j.1365-2435.1997.00040.x.

7. Ruppert, 529–530
8. Ruppert, 531–532
9. Harland, D. P. & Jackson, R. R. (2000). “”Eight-legged cats” and how they see — a review of recent research on jumping spiders (Araneae: Salticidae)” (PDF). Cimbebasia. 16: 231–240. Retrieved 2008-10-11.
10. Wilcox, R. Stimson; Jackson, Robert R. (1998). “Cognitive Abilities of Araneophagic Jumping Spiders”. In Balda, Russell P.; Pepperberg, Irene M.; Kamil, Alan C. Animal cognition in nature: the convergence of psychology and biology in laboratory and field. Academic Press. ISBN 978-0-12-077030-4. Retrieved 2016-05-08.
11. Ruppert, 532–537
12. Ruppert, 578–580
13. Deeleman-Reinhold (2001), p. 27.
14. http://news.discovery.com/animals/spider-detect-movement-senses-111025.htm
15. Barnes, R. S. K., Calow, P., Olive, P., Golding, D., and Spicer, J. (2001). “Invertebrates with Legs: the Arthropods and Similar Groups”. The Invertebrates: A Synthesis. Blackwell Publishing. p. 168. ISBN 0-632-04761-5.
16. Weihmann, Tom; Günther, Michael; Blickhan, Reinhard (2012-02-15). “Hydraulic Leg Extension Is Not Necessarily the Main Drive in Large Spiders”. The Journal of Experimental Biology. 215 (4): 578–583. doi:10.1242/jeb.054585. ISSN 0022-0949. Retrieved 2012-05-21.
17. Parry, D. A. & Brown, R. H. J. (1959). “The Hydraulic Mechanism of the Spider Leg” (PDF). Journal of Experimental Biology. 36 (2): 423–433. Retrieved 2008-09-25.
18. Ruppert, 325–349
19. Vollrath, F. & Knight, D. P. (2001). “Liquid crystalline spinning of spider silk”. Nature. 410 (6828): 541–548. Bibcode:2001Natur.410..541V. doi:10.1038/35069000. PMID 11279484.
20. Tarantulas Shoot Silk from Their Feet
21. Ruppert, 537–539
22. Knoflach, B. & van Harten, A. (2001). “Tidarren argo sp. nov (Araneae: Theridiidae) and its exceptional copulatory behaviour: emasculation, male palpal organ as a mating plug and sexual cannibalism”. Journal of Zoology. 254 (4): 449–459. doi:10.1017/S0952836901000954.
23. Andrade, Maydianne C. B. (2003). “Risky mate search and male self-sacrifice in redback spiders”. Behavioral Ecology. 14 (4): 531–538. doi:10.1093/beheco/arg015.
24. Foelix, R. F. (1996). “Reproduction”. Biology of Spiders. Oxford University Press US. pp. 176–212. ISBN 0-19-509594-4.
25. Ruppert, 523–524
26. Foelix, Rainer F. (1996). Biology of Spiders. Oxford University Press. pp.

232–233. ISBN 0-674-07431-9.

27. Levi, Herbert W. and Levi, Lorna R. (2001) Spiders and their Kin, Golden Press, pp. 20 and 44, ISBN 1582381569
28. Oxford, G. S.; Gillespie, R. G. (1998). "Evolution and Ecology of Spider Coloration". Annual Review of Entomology. 43: 619–643. doi:10.1146/annurev.ento.43.1.619. PMID 15012400.
29. Meehan, C. J., Olson, E. J. and Curry, R. L. (21 August 2008). Exploitation of the Pseudomyrmex–Acacia mutualism by a predominantly vegetarian jumping spider (Bagheera kiplingi). 93rd ESA Annual Meeting. Retrieved 2008-10-10.
30. Jackson, R. R.; Pollard, Simon D.; Nelson, Ximena J.; Edwards, G. B.; Barrion, Alberto T. (2001). "Jumping spiders (Araneae: Salticidae) that feed on nectar" (PDF). J. Zool. Lond. 255: 25–29. doi:10.1017/S095283690100108X.
31. Schütz, D. & Taborsky, M. (2003). "Adaptations to an aquatic life may be responsible for the reversed sexual size dimorphism in the water spider, Argyroneta aquatica" (PDF). Evolutionary Ecology Research. 5 (1): 105–117. Retrieved 2008-10-11.
32. Coddington, J. & Sobrevila, C. (1987). "Web manipulation and two stereotyped attack behaviors in the ogre-faced spider Deinopis spinosus Marx (Araneae, Deinopidae)" (PDF). Journal of Arachnology. 15: 213–225. Retrieved 2008-10-11.
33. Eberhard, W. G. (1977). "Aggressive Chemical Mimicry by a Bolas Spider" (PDF). Science. 198 (4322): 1173–1175. Bibcode:1977Sci...198.1173E. doi:10.1126/science.198.4322.1173. PMID 17818935. Retrieved 2008-10-10.
34. Eberhard, W. G. (1980). "The Natural History and Behavior of the Bolas Spider, Mastophora dizzydeani sp. n. (Araneae)". Psyche. 87 (3–4): 143–170. doi:10.1155/1980/81062. Retrieved 2008-10-10.
35. Yeargan, K. V. & Quate, L. W. (1997). "Adult male bolas spiders retain juvenile hunting tactics". Oecologia. 112 (4): 572–576. doi:10.1007/s004420050347.
36. Wilcox, S. & Jackson, R. (2002). "Jumping Spider Tricksters". In Bekoff, M.; Allen, C. & Burghardt, G. M. The Cognitive Animal: Empirical and Theoretical Perspectives on Animal Cognition (PDF). MIT Press. pp. 27–34. ISBN 0-262-52322-1. Retrieved 25 Mar 2011.
37. Mclver, J. D. & Stonedahl, G. (1993). "Myrmecomorphy: Morphological and Behavioral Mimicry of Ants". Annual Review of Entomology. 38: 351–377. doi:10.1146/annurev.en.38.010193.002031.
38. "Different smiles, single species". University of California Museum of Paleontology. Retrieved 2008-10-10.
39. Cooke, J. A. L., Roth, V. D., and Miller, F. H. "The urticating hairs of theraphosid spiders". American Museum Novitates. American Museum of Natural History (2498). Retrieved 2008-10-11.
40. Blackledge, T. A. & Wenzel, J. W. (2001). "Silk Mediated Defense by an Orb Web Spider against Predatory Mud-dauber Wasps". Behaviour. 138 (2): 155–171. doi:10.1163/15685390151074357.

41. Armstrong, S. (14 July 1990). "Fog, wind and heat — life in the Namib desert". New Scientist. Retrieved 2008-10-11.
42. Vollrath, F. (1986). "Eusociality and extraordinary sex ratios in the spider Anelosimus eximius (Araneae: Theridiidae)". Behavioral Ecology and Sociobiology. 18 (4): 283–287. doi:10.1007/BF00300005.
43. Agnarsson, I. & Kuntner, M. (2005). "Madagascar: an unexpected hotspot of social Anelosimus spider diversity (Araneae: Theridiidae)". Systematic Entomology. 30 (4): 575–592. doi:10.1111/j.1365-3113.2005.00289.x.
44. Avilés, L., Maddison, W. P. and Agnarsson, I. (2006). "A New Independently Derived Social Spider with Explosive Colony Proliferation and a Female Size Dimorphism". Biotropica. 38 (6): 743–753. doi:10.1111/j.1744-7429.2006.00202.x.
45. Matsumoto, T. (1998). "Cooperative prey capture in the communal web spider, Philoponella raffray (Araneae, Uloboridae)" (PDF). Journal of Arachnology. 26: 392–396. Retrieved 2008-10-11.
46. Cangialosi, K. R. (1990). "Social spider defense against kleptoparasitism". Behavioral Ecology and Sociobiology. 27 (1). doi:10.1007/BF00183313.
47. Bertani, R., Fukushima, C. S., and Martins, R. (2008). "Sociable widow spiders? Evidence of subsociality in LatrodectusWalckenaer, 1805 (Araneae, Theridiidae)". Journal of Ethology. 26 (2): 299–302. doi:10.1007/s10164-007-0082-8.
48. Eberhard, W. G. (1990). "Function and Phylogeny of Spider Webs" (PDF). Annual Review of Ecology and Systematics. 21: 341–372. doi:10.1146/annurev.es.21.110190.002013. Retrieved 2008-10-15.
49. Agnarsson, I. (2004). "Morphological phylogeny of cobweb spiders and their relatives (Araneae, Araneoidea, Theridiidae)". Zoological Journal of the Linnean Society. 141 (4): 447–626. doi:10.1111/j.1096-3642.2004.00120.x.
50. Herberstein, M. E. (2000). "Australian Journal of Zoology". Australian Journal of Zoology. 48 (2): 217–223. doi:10.1071/ZO00007.
51. Li, D. & Lee, W. S. (2004). "Predator-induced plasticity in web-building behaviour". Animal Behaviour. 67 (2): 309–318. doi:10.1016/j.anbehav.2003.06.011.
52. Thomson, Peggy & Park, Edwards. "Odd Tales from the Smithsonian". Retrieved 2008-07-21.
53. Schütt, K. (1995). "Drapetisca socialis (Araneae: Linyphiidae): Web reduction — ethological and morphological adaptations" (PDF). European Journal of Entomology. 92: 553–563. Retrieved 2008-10-19.
54. Selden, P. A., Anderson, H. M. and Anderson, J. M. (2009). "A review of the fossil record of spiders (Araneae) with special reference to Africa, and description of a new specimen from the Triassic Molteno Formation of South Africa". African Invertebrates. 50 (1): 105–116. doi:10.5733/afin.050.0103. Abstract PDF
55. Dunlop, Jason A.; David Penney; O. Erik Tetlie; Lyall I. Anderson (2008). "How many species of fossil arachnids are there?". The Journal of Arachnology. 36 (2): 267–272. doi:10.1636/CH07-89.1.
56. Penney, D. & Selden, P. A. (2007). "Spinning with the dinosaurs: the fossil

record of spiders". Geology Today. 23 (6): 231–237. doi:10.1111/j.1365-2451.2007.00641.x.

57. Hecht, H. "Oldest spider web found in amber". New Scientist. Retrieved 2008-10-15.

58. Dunlop, J. A. (1996). "A trigonotarbid arachnid from the Upper Silurian of Shropshire" (PDF). Palaeontology. 39 (3): 605–614. Retrieved 2008-10-12. The fossil was originally named Eotarbus but was renamed when it was realized that a Carboniferous arachnid had already been named Eotarbus: Dunlop, J. A. (1999). "A replacement name for the trigonotarbid arachnid Eotarbus Dunlop". Palaeontology. 42 (1): 191. doi:10.1111/1475-4983.00068.

59. Vollrath, F. & Selden, P. A. (2007). "The Role of Behavior in the Evolution of Spiders, Silks, and Webs" (PDF). Annual Review of Ecology, Evolution, and Systematics. 38: 819–846. doi:10.1146/annurev.ecolsys.37.091305.110221. Retrieved 2008-10-12.

60. Selden, P. A. & Shear, W. A. (December 2008). "Fossil evidence for the origin of spider spinnerets". PNAS. 105 (52): 20781–5. Bibcode:2008PNAS..10520781S. doi:10.1073/pnas.0809174106. PMC 2634869. PMID 19104044.

61. Selden, P. A.; ChungKun Shih; Dong Ren (2011). "A golden orb-weaver spider(Araneae: Nephilidae: Nephila) from the Middle Jurassic of China". Biology Letters. 7 (5): 775–8. doi:10.1098/rsbl.2011.0228. PMC 3169061 . PMID 21508021.

62. Selden, P. A. (1996). "Fossil mesothele spiders". Nature. 379 (6565): 498–499. Bibcode:1996Natur.379..498S. doi:10.1038/379498b0.

63. J. W. Shultz (2007). "A phylogenetic analysis of the arachnid orders based on morphological characters". Zoological Journal of the Linnean Society. 150: 221–265. doi:10.1111/j.1096-3642.2007.00284.x.

64. Shultz, J. W. (2007). "A phylogenetic analysis of the arachnid orders based on morphological characters". Zoological Journal of the Linnean Society. 150 (2): 221–265. doi:10.1111/j.1096-3642.2007.00284.x.

65. Gould, S. J. (1990). Wonderful Life: The Burgess Shale and the Nature of History. Hutchinson Radius. pp. 102–106 . ISBN 0-09-174271-4.

66. Coddington, J. A. (2005). "Phylogeny and Classification of Spiders". In Ubick, D.; Paquin, P.; Cushing, P. E.; Roth, V. Spiders of North America: an identification manual (PDF). American Arachnological Society. pp. 18–24. ISBN 0-9771439-0-2. Retrieved 2008-10-12.

67. Leroy, J & Leroy, A. (2003). "How spiders function". Spiders of Southern Africa. Struik. pp. 15–21. ISBN 1-86872-944-3.

68. Ono, H. (2002). "New and Remarkable Spiders of the Families Liphistiidae, Argyronetidae, Pisauridae, Theridiidae and Araneidae (Arachnida) from Japan". Bulletin of the National Science Museum (of Japan), Series A. 28 (1): 51–60.

69. Coyle, F. A. (1986). "The Role of Silk in Prey Capture". In Shear, W. A. Spiders—webs, behavior, and evolution. Stanford University Press. pp. 272–273.

ISBN 0-8047-1203-4.

70. Forster, R. R. & Platnick, N. I. (1984). "A review of the archaeid spiders and their relatives, with notes on the limits of the superfamily Palpimanoidea (Arachnida, Araneae)" (abstract). Bulletin of the American Museum of Natural History. 178: 1–106. Retrieved 2008-10-13. Full text at "A review of the archaeid spiders and their relatives" (PDF). Retrieved 2008-10-13. (60 MB)
71. Penney, D. & Selden, P. A. Deltshev, C. & Stoev, P., eds. "European Arachnology 2005" (PDF). Acta Zoologica Bulgarica. Supplement No. 1: 25–39. Retrieved 2008-10-13. |contribution= ignored (help)
72. "Natural history of Mygalomorphae". Agricultural Research Council of New Zealand. Retrieved 2008-10-13.
73. Foelix, Rainer F. Biology of Spiders (2nd ed.). New York: Oxford University Press. pp. 4–5. ISBN 0-19-973482-8.
74. Spider Bite Symptoms and First Aid By Rod Brouhard, About.com. Updated: October 19, 2008
75. Vetter, Richard S.; Isbister, Geoffrey K. (2008). "Medical Aspects of Spider Bites". Annual Review of Entomology. 53: 409–29. doi:10.1146/annurev.ento.53.103106.093503. PMID 17877450.
76. "Spiders". Illinois Department of Public Health. Retrieved 2008-10-11.
77. Vetter RS, Barger DK (2002). "An infestation of 2,055 brown recluse spiders (Araneae: Sicariidae) and no envenomations in a Kansas home: implications for bite diagnoses in nonendemic areas". Journal of Medical Entomology. 39 (6): 948–51. doi:10.1603/0022-2585-39.6.948. PMID 12495200.
78. Hannum, C. & Miller, D. M. "Widow Spiders". Department of Entomology, Virginia Tech. Retrieved 2008-10-11.
79. "Funnel web spiders". Australian Venom Research Unit. Retrieved 2008-10-11.
80. "Pub chef bitten by deadly spider". BBC. 2005-04-27. Retrieved 2008-10-11.
81. Diaz, J. H. (August 1, 2004). "The Global Epidemiology, Syndromic Classification, Management, and Prevention of Spider Bites". American Journal of Tropical Medicine and Hygiene. 71 (2): 239–250. PMID 15306718.
82. Williamson, J. A., Fenner, P. J., Burnett, J. W., and Rifkin, J. (1996). Venomous and Poisonous Marine Animals: A Medical and Biological Handbook. UNSW Press. pp. 65–8. ISBN 0-86840-279-6.
83. Nishioka, S de A. (2001). "Misdiagnosis of brown recluse spider bite". Western Journal of Medicine. 174 (4): 240. doi:10.1136/ewjm.174.4.240. PMC 1071344 . PMID 11290673.
84. Isbister GK (2001). "Spider mythology across the world". Western Journal of Medicine. 174 (4): 86–7. doi:10.1136/ewjm.175.2.86. PMC 1071491 . PMID 11483545.
85. "Spider Venom Could Yield Eco-Friendly Insecticides". National Science Foundation (USA). Retrieved 2008-10-11.

86. Salima Machkour M'Rabet, Yann Hénaut, Peter Winterton and Roberto Rojo (2011). "A case of zootherapy with the tarantula Brachypelma vagans Ausserer, 1875 in traditional medicine of the Chol Mayan ethnic group in Mexico". Journal of Ethnobiology and Ethnomedicine.

87. Novak, K. (2001). "Spider venom helps hearts keep their rhythm". Nature Medicine. 7 (155): 155. doi:10.1038/84588. PMID 11175840.

88. Lewis, R. J. & Garcia, M. L. (2003). "Therapeutic potential of venom peptides" (PDF). Nature Reviews Drug Discovery. 2 (10): 790–802. doi:10.1038/nrd1197. PMID 14526382. Retrieved 2008-10-11.

89. Bogin, O. (Spring 2005). "Venom Peptides and their Mimetics as Potential Drugs" (PDF). Modulator (19). Retrieved 2008-10-11.

90. Andrade E; Villanova F; Borra P; Leite, Katia; Troncone, Lanfranco; Cortez, Italo; Messina, Leonardo; Paranhos, Mario; et al. (2008). "Penile erection induced in vivo by a purified toxin from the Brazilian spider Phoneutria nigriventer". British Journal of Urology International. 102 (7): 835–7. doi:10.1111/j.1464-410X.2008.07762.x. PMID 18537953.

91. Salima Machkour-M'Rabet, Yann Hénaut, Peter Winterton and Roberto Rojo (2011). "A case of zootherapy with the tarantula Brachypelma vagans Ausserer, 1875 in traditional medicine of the Chol Mayan ethnic group in Mexico". Journal of ethnobiology and ethno medicine.

92. Hinman, M. B., Jones J. A., and Lewis, R. W. (2000). "Synthetic spider silk: a modular fiber" (PDF). Trends in Biotechnology. 18 (9): 374–9. doi:10.1016/S0167-7799(00)01481-5. PMID 10942961. Retrieved 2008-10-19.

93. Menassa, R., Zhu, H., Karatzas, C. N., Lazaris, A., Richman, A., and Brandle, J. (2004). "Spider dragline silk proteins in transgenic tobacco leaves: accumulation and field production". Plant Biotechnology Journal. 2 (5): 431–8. doi:10.1111/j.1467-7652.2004.00087.x. PMID 17168889.

94. Ray, N. (2002). Lonely Planet Cambodia. Lonely Planet Publications. p. 308. ISBN 1-74059-111-9.

95. Weil, C. (2006). Fierce Food. Plume. ISBN 0-452-28700-6. Retrieved 2008-10-03.

96. "A Common Phobia". phobias-help.com. Archived from the original on 2009-08-02. Retrieved 2009-08-02. There are many common phobias, but surprisingly, the most common phobia is arachnophobia.

97. Fritscher, Lisa (2009-06-03). "Spider Fears or Arachnophobia". Phobias. About.com. Archived from the original on 2009-08-02. Retrieved 2009-08-02. Arachnophobia, or fear of spiders, is one of the most common specific phobias.

98. "The 10 Most Common Phobias — Did You Know?". 10 Most Common Phobias. Archived from the original on 2009-08-02. Retrieved 2009-08-02. Probably the most recognized of the 10 most common phobias, arachnophobia is the fear of spiders. The statistics clearly show that more than 50% of women and 10% of men show signs of this leader on the 10 most common phobias list.

99. Friedenberg, J. & Silverman, G. (2005). Cognitive Science: An Introduction to

the Study of Mind. SAGE. pp. 244–245. ISBN 1-4129-2568-1.

100. Davey, G. C. L. (1994). "The "Disgusting" Spider: The Role of Disease and Illness in the Perpetuation of Fear of Spiders". Society and Animals. 2 (1): 17–25. doi:10.1163/156853094X00045.
101. De Vos, Gail (1996). Tales, Rumors, and Gossip: Exploring Contemporary Folk Literature in Grades 7–12. Libraries Unlimited. p. 186. ISBN 1-56308-190-3.
102. Garai, Jana (1973). The Book of Symbols. New York: Simon & Schuster. ISBN 0-671-21773-9.
103. De Laguna, Frederica (2002). American Anthropology: Papers from the American Anthropologist. University of Nebraska Press. p. 455. ISBN 0-8032-8280-X.
104. Benson, Elizabeth. The Mochica: A Culture of Peru. New York: Praeger Press. 1972.
105. Berrin, Katherine & Larco Museum. The Spirit of Ancient Peru: Treasures from the Museo Arqueológico Rafael Larco Herrera. New York: Thames and Hudson, 1997.

CHAPTER 3

Ethnoentomology

Chapter Contents

- **Introduction**
- **Context**
- **Practical Uses**

Humans spend their intellectual energies in three basic areas of activity: surviving, using practical learning (the application of technology); seeking pure knowledge through inductive mental processes (science); and pursuing enlightenment to taste a pleasure by aesthetic exercises that may be referred to as the "humanities." Entomology has long been concerned with survival (economic or applied entomology) and scientific study (academic entomology), but the branch of investigation that addresses the influence of insects (and other terrestrial Arthropoda, including arachnids, myriapods, etc) in literature, language, music, the arts, interpretive history, religion, and recreation has only recently been recognized as a distinct field. This is referred to as "cultural entomology".

INTRODUCTION

The "Spanish Fly", *Lytta vesicatoria* has been considered to have medicinal, aphrodisiac, and other properties.

The roles of **insects in culture** span different aspects of human life, whether analysed academically or more generally.

Academically, the interaction of insects and society has been treated in part as **cultural entomology**, dealing mostly with "advanced" societies, and in part as **ethnoentomology**, dealing mostly with "primitive" societies, though the distinction is weak and not based on theory. Both academic disciplines explore the parallels, connections and influence of insects on human populations, and vice versa. They are rooted in anthropology and natural history, as well as entomology, the study of insects. Other cultural uses of insects, such as biomimicry, do not necessarily lie within these academic disciplines.

More generally, people make a wide range of uses of insects, both practical and symbolic.

Practical uses include as food (entomophagy), in medicine, for the valuable textile silk, for dyestuffs such as carmine, in science, where the fruit fly is an important model organism in genetics, and in warfare, where insects were successfully used in the Second World War to spread disease in enemy populations. One insect, the honey bee, provides honey, pollen, royal jelly, propolis and an anti-inflammatory peptide, melittin; its larvae too are eaten in some societies. Medical uses of insects include maggot therapy for wound debridement. Over a thousand protein families have been identified in the saliva of blood-feeding insects; these may provide useful drugs such as anticoagulants, vasodilators, antihistamines and anaesthetics.

Symbolic uses include roles in art, in music (with many songs featuring insects), in film, in literature, in religion, and in mythology. Insect costumes are used in theatrical productions and worn for parties and carnivals.

CONTEXT

Cultural entomology and Ethnoentomology

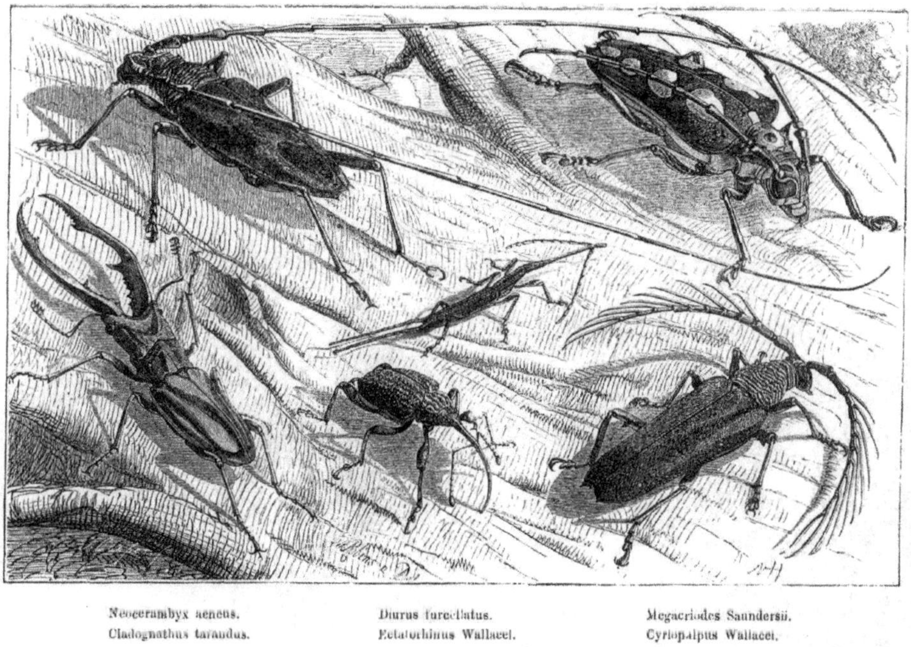

"Remarkable Beetles Found at Simunjon, Borneo": Alfred Russel Wallace was a pioneer of ethnoentomology.

Ethnoentomology developed from the 19th century with early works by authors such as Alfred Russel Wallace (1852) and Henry Walter Bates (1862). Hans Zinsser's classic *Rats, Lice and History* (1935) showed that insects were an important force in human history. Writers like William Morton Wheeler, Maurice Maeterlinck, and Jean Henri Fabre described insect life and communicated their meaning to people "with imagination and brilliance". Frederick Simon Bodenheimer's *Insects as Human Food* (1951) drew attention to the scope and potential of entomophagy, and showed a positive aspect of insects. Food is the most studied topic in ethnoentomology, followed by medicine and beekeeping.

In 1968, Erwin Schimitschek (de) claimed cultural entomology as a branch of insect studies, in a review of the roles insects played in folklore and culture including religion, food, medicine and the arts. In 1984, Charles Hogue covered the field in English and from 1994 to 1997, Hogue's *The Cultural Entomology Digest* served as a forum on the field. Hogue argued that "Humans spend their intellectual energies in three basic areas of activity: surviving, using practical learning (the application of technology); seeking pure knowledge through inductive mental processes (science); and pursuing enlightenment to taste a pleasure by aesthetic exercises that may be referred to as the "humanities." Entomology has long been concerned with survival (economic entomology) and scientific study (academic entomology), but the branch of investigation that addresses the influence of insects (and other terrestrial Arthropoda, including arachnids and myriapods) in literature, language, music, the arts, interpretive history, religion, and recreation has only become recognized as a distinct field through Schimitschek's work. Hogue set out the boundaries of the field by saying: The narrative history of the science of entomology is not part of cultural entomology, while the influence of insects on general history would be considered cultural entomology." He added: "Because the term "cultural" is narrowly defined, some aspects normally included in studies of human societies are excluded."

Darrell Addison Posey, noting that the boundary between cultural entomology and ethnoentomology is difficult to draw, cites Hogue as limiting cultural entomology to the influence of insects on the essence of humanity as expressed in the arts and humanities. Posey notes further that cultural anthropology is usually restricted to the study of advanced, industrialised, and literate societies, whereas ethnoentomology studies the entomological concerns of ‹primitive' or ‹noncivilized' societies. Posey states at once that the division is artificial, complete with an unjustified us/them bias. Brian Morris similarly criticises the way that anthropologists treat non-Western attitudes to nature as monadic and spiritualist, and contrast this in gnostic fashion" with a simplistic treatment of Western, often 17th century, mechanistic attitude. Morris considers this "quite unhelpful, if not misleading", and offers instead his own research into the multiple ways that the people of Malawi relate to insects and other animals: "pragmatic, intellectual, realist, practical, aesthetic, symbolic and sacramental."

Insect Ecosystem Services

Pollination, here by a bee on an avocado crop, is an ecosystem service.

The Millennium Ecosystem Assessment (MEA) report 2005 defines ecosystem services as benefits people obtain from ecosystems, and distinguishes four categories, namely provisioning, regulating, supporting, and cultural. A fundamental tenet is that a few species of arthropod are well understood for their influence on humans (such as honeybees, ants, mosquitoes, and spiders). However, insects offer ecological goods and services. The Xerces Society calculates the economic impact of four ecological services rendered by insects: pollination, recreation (i.e. "the importance of bugs to hunting, fishing, and wildlife observation, including bird-watching), dung burial, and pest control. The value has been estimated at $153 billion worldwide. As the ant expert E. O. Wilson observed: "If all mankind were to disappear, the world would regenerate back to the rich state of equilibrium that existed ten thousand years ago. If insects were to vanish, the environment would collapse into chaos." A Nova (TV Series) segment on the American Public Broadcasting Service framed the relationship with insects in an urban context: "We humans like to think that we run the world. But even in the heart of our great cities, a rival superpower thrives ... These tiny creatures live all around us in vast numbers, though we hardly even notice them. But in many ways, it is they who really run the show. *The Washington Post* stated: "We are flying blind in many aspects of preserving the environment, and that's why we are so surprised when a species like the honeybee starts to crash, or an insect we don't want, the Asian tiger mosquito or the fire ant, appears in our midst. In other words: Start thinking about the bugs.

Attitudes to Insects

Steven Kellert classified public attitudes towards insects by 9 primary interests, namely aesthetic, humanistic, moralistic, naturalistic, dominionistic, ecologistic, negativistic, utilitarian, and scientific.

Human attitudes toward insects are often negative, reinforced by sensationalism in the media. This has produced a society that attempts to eliminate insects from daily life. Nearly 75 million pounds of broad-spectrum insecticides are manufactured and sold each year for use in American homes and gardens. Annual revenues from insecticide sales to homeowners exceeded $450 million in 2004. Out of the roughly a million species of insects described so far, not more than 1,000 (about 1/10 of 1%) can be regarded as serious pests, and less than 10,000 (about 1%) are even occasional or sporadic pests. Yet not one species of insect has been permanently eradicated through the use of pesticides. Instead, at least 1,000 species have developed field resistance to pesticides.

PRACTICAL USES

As Food

Entomophagy is the eating of insects. Many insects are considered a culinary delicacy in some societies around the world, and Frederick Simon Bodenheimer's *Insects as Human Food* (1951) drew attention to the scope and potential of entomophagy, but the practice is uncommon and even taboo in other societies. Sometimes insects are considered suitable only for the poor in the third world, but in 1975 Victor Meyer-Rochow suggested that insects could help ease global future food shortages and advocated a change in western attitudes towards cultures in which insects were appreciated as a food item. P.J. Gullan and P.S. Cranston felt that the remedy for this may be marketing of insect dishes as suitably exotic and costly to make them acceptable. They also note that some societies in sub-Saharan Africa prefer caterpillars to beef, while Chakravorty et al. (2011) point out that food insects (highly appreciated in North-East India) are more expensive than meat. The economics, i.e., the costs involved collecting food insects and the money earned through the sale of such insects, have been studied in a Laotian setting by Meyer-Rochow et al. (2008). In Mexico, ant larvae and Corixid water boatman eggs are sought out as a form of caviar by gastronomes. In Guangdong, water beetles fetch a high enough price for these insects to be farmed. Especially high prices are fetched in Thailand for the giant water bug *Lethocerus indicus*.

Witchetty grubs were prized as food by Australian aborigines.

Insects used in food include honey bee larvae and pupae, mopani worms, silkworms, Maguey worms, Witchetty grubscrickets, grasshoppers and locusts. In Thailand, there are 20,000 farmers rearing crickets, producing some 7,500 tons per year.

In Medicine

Insects have been used medicinally in cultures around the world, often according to the Doctrine of Signatures. Thus, the femurs of grasshoppers, which were said to resemble the human liver, were used to treat liver ailments by the indigenous peoples of Mexico. The doctrine was applied in both Traditional Chinese Medicine (TCM) and in Ayurveda. TCM uses arthropods for various purposes; for example, centipede is used to treat tetanus, seizures, and convulsions, while the Chinese Black Mountain Ant, *Polyrhachis vicina*, is used as a cure all, especially by the elderly, and extracts have been examined as a possible anti-cancer agent. Ayurveda uses insects such as Termite for conditions such as ulcers, rheumatic diseases, anaemia, and pain. The Jatropha leaf miner's larvae are used boiled to induce lactation, reduce fever, and soothe the gastrointestinal tract. In contrast, the traditional insect medicine of Africa is local and unformalised. The indigenous peoples of Central America used a wide variety of insects medicinally. Mayans used Army ant soldiers as living sutures. The venom of the Red harvester ant was used to cure rheumatism, arthritis, and poliomyelitis via the immune reaction produced by its sting. Boiled *silkworm* pupae were taken to treat apoplexy, aphasy, bronchitis, pneumonia, convulsions, haemorrhages, and frequent urination.

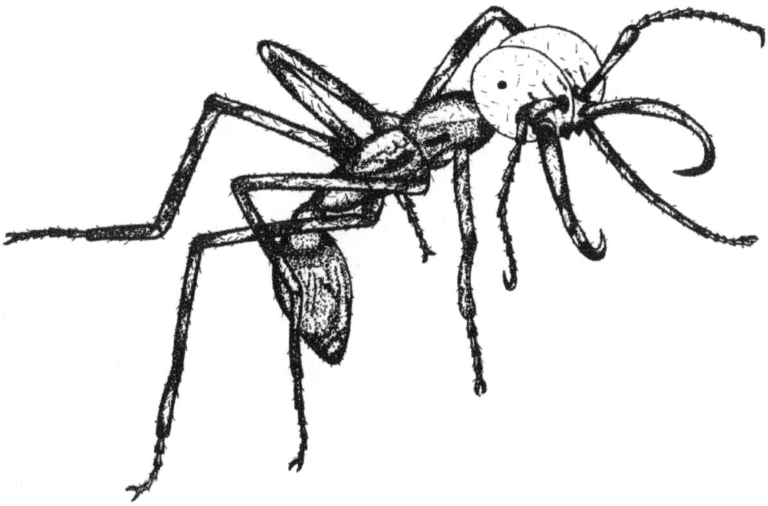

Army ants were used by the Mayans as living sutures, their powerful jaws holding a wound closed.

Honey bee products are used medicinally in apitherapy across Asia, Europe, Africa, Australia, and the Americas, despite the fact that the honey bee was not introduced to the Americas until the colonization by Spain and Portugal. They are by far the most common medical insect product both historically and currently, and the most frequently referenced of these is honey. It can be applied to skin to treat excessive scar tissue, rashes, and burns, and as an eye poultice to treat infection. Honey is taken for digestive problems and as a general health restorative. It is taken hot to treat colds, cough, throat infections, laryngitis, tuberculosis, and lung diseases. Apitoxin (honey bee venom) is applied via direct stings to relieve arthritis, rheumatism, polyneuritis, and asthma. Propolis, a resinous, waxy mixture collected by honeybees and used as a hive insulator and sealant, is often consumed by menopausal women because of its high hormone content, and it is said to have antibiotic, anesthetic, and anti-inflammatory properties. Royal jelly is used to treat anaemia, gastrointestinal ulcers, arteriosclerosis, hypo- and hypertension, and inhibition of sexual libido. Finally Bee bread, or bee pollen, is eaten as a generally health restorative, and is said to help treat both internal and external infections. One of the major peptides in bee venom, Melittin, has the potential to treat inflammation in sufferers of Rheumatoid arthritis and Multiple sclerosis.

The rise of antibiotic resistant infections has sparked pharmaceutical research for new resources, including into arthropods.

Maggot therapy uses blowfly larvae to perform wound-cleaning debridement. They secrete allantoin, which is used to treat the infectious bone disease, Osteomyelitis.

Cantharidin, the blister-causing oil found in several families of beetles described by the vague common name Spanish fly, was accepted by the FDA in 2004 as treatment for warts and other skin problems. It was in Ancient Greece and Rome, and has been used as an aphrodisiac in some societies. Studies in cell culture and animal models have demonstrated tumour-fighting properties.

Blood-feeding insects like ticks, horseflies, and mosquitoes inject multiple bioactive compounds into their prey. These insects have long been used by practitioners of Eastern Medicine to prevent blood clot formation or thrombosis, suggesting possible applications in scientific medicine. Over 1280 protein families have been associated with the saliva of blood feeding organisms, including inhibitors of platelet aggregation, ADP, arachidonic acid, thrombin, PAF, anticoagulants, vasodilators, vasoconstrictors, antihistamines, sodium channel blockers, complement inhibitors, pore formers, inhibitors of angiogenesis, anaesthetics, AMPs and microbial pattern recognition molecules, and parasite enhancers/activators. Ixolaris, a tissue factor inhibitor has been shown to block primary tumour growth and angiogenesis in a glioblastoma model.

Arachnids have been used in traditional medicine, and their venom has been studied for bioactive factors. In 1993 Margatoxin was synthesized from the venom of the Central American bark scorpion, *Centruroides margaritatus*. This peptide selectively inhibits voltage-gated potassium channels. Patented by Merck, it has the potential to prevent neointimal hyperplasia, a common cause of bypass graft failure.

In Science and Technology

The common fruitfly *Drosophila melanogaster* is one of the most widely used model organisms in biological research.

Insects play an important role in biological research. Because of its small size, short generation time and high fecundity, the common fruit fly *Drosophila melanogaster* was selected as a model organism for studies of the genetics of higher eukaryotes. *D. melanogaster* has been an essential part of studies into principles like genetic linkage, interactions between genes, chromosomal genetics, evolutionary developmental biology, animal behaviour and evolution. Because genetic systems are well conserved among eukaryotes, understanding basic cellular processes like DNA replication or transcription in fruit flies helps scientists to understand those processes in other eukaryotes, including humans. The genome of *D. melanogaster* was sequenced in 2000, reflecting the fruit fly's important role in biological research. 70% of the fly genome is similar to the human genome, supporting the Darwinian theory of evolution from a single origin of life.

Cochineal scale insects being collected from a prickly pear in Central America. Illustration by José Antonio de Alzate y Ramírez, 1777.

Some hemipterans are used to produce dyestuffs such as carmine (also called cochineal). The scale insect *Dactylopius coccus* produces the brilliant red-coloured carminic acid to deter predators. Up to 100,000 scale insects are needed to make a kilogram (2.2 lbs) of cochineal dye.

A similarly enormous number of lac bugs are needed to make a kilogram of shellac, a brush-on colourant and wood finish. Additional uses of this traditional product include the waxing of citrus fruits to extend their shelf-life, and the coating of pills to moisture-proof them, provide slow-release or mask the taste of bitter ingredients.

Kermes is a red dye from the dried bodies of the females of a scale insect in the genus *Kermes*, primarily *Kermes vermilio*. *Kermes* are native to the Mediterranean region, living on the sap of the Kermes oak. They were used as a red dye by the ancient Greeks and Romans. The kermes dye is a rich red, and has good colour fastness in silk and wool.

Target for biomimicry research: the Namib desert beetle, *Stenocara gracilipes*, channels water from fog down its wings.

Insect attributes are sometimes mimicked in architecture, as at the Eastgate Centre, Harare, which uses passive cooling, storing heat in the morning and releasing it in the warm parts of the day. The target of this piece of biomimicry is the structure of the mounds of termites such as *Macrotermes michaelseni* which effectively cool the nests of these social insects. The properties of the Namib desert beetle's exoskeleton, in particular its wing-cases (elytra) which have bumps with hydrophilic (water-attracting) tips and hydrophobic (water-shedding) sides, have been mimicked in a film coating designed for the British Ministry of Defence, to capture water in arid regions.

In Textiles

Silkworms, the caterpillars and pupae of the moth *Bombyx mori*, have been reared to produce silk in China from the Neolithic Yangshao period onwards, c. 5000 BC. Production spread to India by 140 AD. The caterpillars are fed on mulberry leaves. The cocoon, produced after the fourth moult, is covered with a continuous filament of the silk protein, fibroin, gummed together with sericin. In the traditional process, the gum is removed by soaking in hot water, and the silk is then unwound from the cocoon and reeled. Filaments are spun together to make silk thread. Commerce in silk between China and countries to its west began in ancient times, with silk known from an Egyptian mummy of 1070 BC, and later to the ancient Greeks and Romans. The silk road leading west from China was opened in the second century AD, helping to drive trade in silk and other goods.

In Warfare

The use of insects for warfare may have been attempted in the Middle Ages or earlier, but was first systematically researched by several nations during the twentieth century. It was put into practice by the Japanese army's Unit 731 in attacks on China during the Second World War, killing almost 500,000 Chinese people with fleas infected with plague and flies infected with cholera. Also in the Second World War, the Germans explored the use of Colorado beetles to destroy enemy potato crops. During the Cold War, the US Army considered using yellow fever mosquitoes to attack Soviet cities.

SYMBOLIC USES

In mythology and Folklore

Gold plaques embossed with winged bee goddesses, perhaps the Thriai, found at Camiros Rhodes, dated to the 7th century BC.

Insects have appeared in mythology around the world from ancient times. Among the insect groups featuring in myths are the bee, butterfly, cicada, dragonfly, praying mantis and scarab beetle. Scarab beetles held religious and cultural symbolism in Old Egypt, Greece and some shamanistic Old World cultures. The ancient Chinese regarded cicadas as symbols of rebirth or immortality. In Mesopotamian literature, the epic poem of Gilgamesh has allusions to dragonflies, signifying the impossibility of immortality. Amongst the Arrernte Aborigines of Australia, honey ants and witchety grubs served as personal clan totems. In the case of the San bushmen of the Kalahari, it is the praying mantis which holds much cultural significance including creation and zen-like patience in waiting.

Insects feature in folklore around the world. In China, farmers traditionally regulated their crop planting according to the Awakening of the Insects, when temperature shifts and monsoon rains bring insects out of hibernation. Most "Awakening" customs are related to eating snacks like pancakes, parched beans, pears, and fried corn, symbolizing harmful insects in the field.

Woolly bear caterpillar larva of the Isabella tiger moth, *Pyrrharctia isabella* is celebrated in the annual Woollybear Festival of the American Great Lakes.

In the Great Lakes region of the United States, there is an annual Woollybear Festival that has been celebrated for over 40 years. The larvae of the species *Pyrrharctia isabella* (commonly known as the Isabella Tiger moth), with their 13 distinct segments of black and reddish-brown, have the reputation in common folklore of being able to forecast the coming winter weather. From 1948, C. H. Curran, curator of insects at the American Museum of Natural History, counted the caterpillars in Bear Mountain State Park in an attempt to validate that tradition.

During the 20th Century and into the new millennium, many misconceptions of insects – both positive and negative – have been dispelled. Karl von Frisch worked on insects' stimulus perception and communication. Much to the skepticism of his contemporaries, Von Frisch was the first scientist to demonstrate that invertebrates could see colors. Until the 1930s, colored sight was thought to be the domain of vertebrates. Von Frisch later proved that honeybees communicated through a definite dance language. Again, this discovery resonated through the natural sciences and anthropology: it made for a compelling re-evaluation of the concept of language, thought to be only the realm of humans and a select group of mammals. The discovery was such a revelation that in 1973 Karl von Frisch received the Nobel Prize for Physiology or Medicine (together with Konrad Lorenz and Nikolaas Tinbergen) for his achievements in comparative behavioral physiology and pioneering work in communication between insects. Ever since the discovery the topic has been a source of controversy: it directly impacts the notion of the intellectual superiority of humans. The anthropologist Hugh Raffles writes of the discovery:

To give them language was to simultaneously celebrate their difference and to doom them to impossibility, to condemn them to the merely imitative ... What foolishness to judge insects – so ancient, so diverse, so accomplished, so successful, so beautiful, so astonishing, so mysterious, so unknown – by criteria they can never meet and about which they could not care. What silliness to disregard their accomplishments and focus instead on their supposed deficiencies!

There is a common misconception that cockroaches are serious vectors of disease. Cockroaches are not vagile, i.e. they do not travel far from their birthplace as opposed to the housefly which may travel miles from its birthplace. The cockroach will spread its GI tract contents throughout a building. If that is a home it will be helping to spread the diseases that the family and friends bring into the home. In a hospital or a restaurant it can spread diseases from sick to well individuals.They have no bite or sting. It is true that by external contact that cockroaches can carry bacteria such as salmonella onto surfaces. Their shells contain a protein, arylphorin, implicated in asthma and other respiratory conditions.

Shoen Uemura - firefly, a sign of summer in Japan.

Many people believe the urban myth that the daddy longlegs Opiliones have the most poisonous bite in the spider world, but that the fangs are too small to penetrate human skin. This is untrue on several counts. None of the known species of harvestmen have venom glands; their chelicerae are not hollowed fangs but grasping claws that are typically very small and definitely not strong enough to break human skin.

In Japan, the emergence of fireflies and rhinoceros beetles signify the anticipated changing of the seasons.

In Religion

Scarab with separate wings, c. 712-342 BC

In the Brazilian Amazon, members of the Tupí–Guaraní language family have been observed using *Pachycondyla commutata* ants during female rite-of-passage ceremonies, and prescribing the sting of *Pseudomyrmex* spp. for fevers and headaches.

The red harvester ant *Pogonomyrmex californicus* has been widely used by natives of Southern California and Northern Mexico for hundreds of years in ceremonies conducted to help tribe members acquire spirit helpers through hallucination. During the ritual, young men are sent away from the tribe and consume large quantities of live, unmasticated ants under the supervision of an elderly member of the tribe. Ingestion of ants should lead to a prolonged state of unconsciousness, where dream helpers appear and serve as allies to the dreamer for the rest of his life.

In Art

Both the symbolic form and the actual body of insects have been used to adorn humans in ancient and modern times. A recurrent theme for ancient cultures in Europe and the Near East regarded the sacred image of a bee or human with insect features. Often referred to as the bee goddess, these images were found in gems and stones. An onyx gem from Knossos (ancient Crete) dating to approximately 1500 BC illustrates a Bee goddess with bull horns above her head. In this instance, the figure is surrounded by dogs with wings, most likely representing Hecate and Artemis – gods of the underworld, similar to the Egyptian gods Akeu and Anubis.

Beetlewing art is an ancient craft technique using iridescent beetle wings practiced traditionally in Thailand, Myanmar, India, China and Japan. Beetlewing pieces are used as an adornment to paintings, textiles and jewelry. Different species of metallic wood-boring beetle wings were used depending on the region, but traditionally the most valued were those from beetles belonging to the genus Sternocera. The practice comes from across Asia and Southeast Asia, especially Thailand, Myanmar, Japan, India and China.

In Thailand beetlewings were preferred to decorate clothing (shawls and Sabai cloth) and jewellery in former court circles.

Radha and Krishna in Rasamanjari by Bhanudatta, Basohli, ca. 1670. Opaque watercolour and gold on paper, with applied beetlewing fragments.

A Dragon-fly, Two Moths, a Spider and Some Beetles, With Wild Strawberries by Jan van Kessel, 17th century. Wasp beetle, top left; clouded border moth, top right; migrant hawker dragonfly and cardinal beetle, centre left; magpie moth, centre right; cockchafer, lower left.

The Canadian entomologist C.H. Curran's 1945 book, *Insects of the Pacific World*, noted women from India and Sri Lanka, who kept 1 1/2 inch long, iridescent greenish coppery beetles of the species *Chrysochroa ocellata* as pets. These living jewels were worn on festive occasions, probably with a small chain attached to one leg anchored to the clothing to prevent escape. Afterwards, the insects were bathed, fed, and housed in decorative cages. Living jewelled beetles have also been worn and kept as pets in Mexico.

Butterflies have long inspired humans with their life cycle, color, and ornate patterns. The novelist Vladimir Nabokov was also a renowned butterfly expert. He published and illustrated many butterfly species, stating:

I discovered in nature the nonutilitarian delights that I sought in art. Both were a form of magic, both were games of intricate enchantment and deception.

It was the aesthetic complexity of insects that led Nabokov to reject natural selection.

The naturalist Ian MacRae writes of butterflies:

... the animal is at once awkward, flimsy, strange, bouncy in flight, yet beautiful and immensely sympathetic; it is painfully transient, albeit capable of extreme migrations and transformations. Images and phrases such as kaleidoscopic instabilities, oxymoron of similarities, rebellious rainbows, visible darkness and souls of stone have much in common. They bring together the two terms of a conceptual contradiction, thereby facilitating the mixing of what should be discrete and mutually exclusive categories ... In positing such questions, butterfly science, an inexhaustible, complex, and finely nuanced field, becomes not unlike the human imagination, or the field of literature itself. In the natural history of the animal, we begin to sense its literary and artistic possibilities.

The photographer Kjell Sandved spent 25 years documenting all 26 characters of the Latin alphabet using the wing patterns of butterflies and moths as *The Butterfly Alphabet*.

In 2011, the artist Anna Collette created over 10,000 individual ceramic insects at Nottingham Castle, "Stirring the Swarm." Reviews of the exhibit offered a compelling narrative for Cultural Entomology: "the unexpected use of materials, dark overtones, and the straightforward impact of thousands of tiny multiples within the space. The exhibition was at once both exquisitely beautiful and deeply repulsive, and this strange duality was fascinating."

In Literature and Film

The Ancient Greek playwright Aeschylus has a gadfly pursue and torment Io, a maiden associated with the moon, watched constantly by the eyes of the herdsman Argus, associated with all the stars: Io: Ah! Hah! Again the prick, the stab of gadfly-sting! O earth, earth, hide, the hollow shape—Argus—that evil thing—the hundred-eyed. William Shakespeare, inspired by Aeschylus, has Tom o'Bedlam in *King Lear*, "Whom the foul fiend hath led through fire and through flame, through ford and whirlpool, o'er bog and quagmire", driven mad by the constant pursuit. In *Antony and Cleopatra*, Shakespeare similarly likens Cleopatra's hasty departure from the Actium battlefield to that of a cow chased by a gadfly. H. G. Wells introduced giant wasps in his 1904 novel *The Food of the Gods and How It Came to Earth*, making use of the newly discovered growth hormones to lend plausibility to his science fiction. Lafcadio Hearn's essay *Butterflies* analyses the treatment of the butterfly in Japanese literature, both prose and poetry. He notes that these often allude to Chinese tales, such as of the young woman that the butterflies took to be a flower. He translates 22 Japanese Haiku poems about butterflies, including one by the Haiku master Matsuo Bashō, said to suggest happiness in springtime: Wake up! Wake up!—I will make thee my comrade, thou sleeping butterfly.

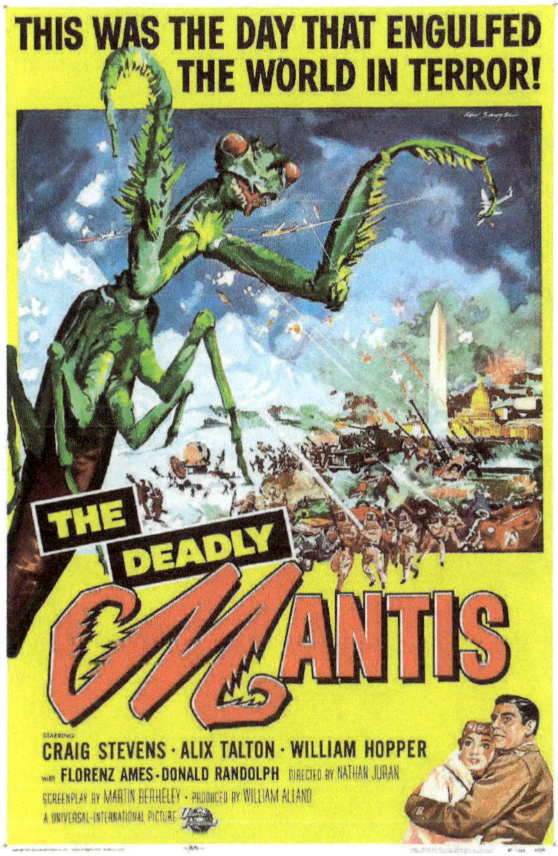

Poster for the 1957 film *The Deadly Mantis*.

The novelist Vladimir Nabokov was the son of a professional lepidopterist, and was interested in butterflies himself. He wrote his novel *Lolita* while travelling on his annual butterfly-collection trips in the western United States. He eventually became a leading lepidopterist. This is reflected in his fiction, where for example *The Gift* devotes two whole chapters (of five) to the tale of a father and son on a butterfly expedition.

Horror films involving insects, sometimes called big bug movies, include the pioneering 1954 *Them!*, featuring giant ants mutated by radiation, and the 1957 *The Deadly Mantis*.

The Far Side, a newspaper cartoon, has been used by professor of Michael Burgett as a teaching tool in his entomology class; *The Far Side* and its author Gary Larson have been acknowledged by biologist Dale H. Clayton his colleague for "the enormous contribution" Larson has made to their field through his cartoons.

In Music

The *Flight of the Bumblebee* was written by Nikolai Rimsky-Korsakov, in 1899–1900, as part of his opera, *The Tale of Tsar of Saltan*. The piece is one of the most recognizable pieces in classical composition. The bumblebee in the story is a prince who has been transformed into an insect so that he can fly off to visit his father. The play upon which the opera was based – written by Alexandr Pushkin – originally had two more insect themes: the Flight of the Mosquito and the Flight of the Fly.

The Hungarian composer Béla Bartók explained in his diary that he was attempting to depict the desperate attempts to escape of a fly caught in a cobweb in his piece *From the Diary of a Fly, for piano (Mikrokosmos Vol. 6/142)*.

The jazz musician and philosophy professor David Rothenberg plays duets with singing insects including cicadas, crickets, and beetles.

In Astronomy

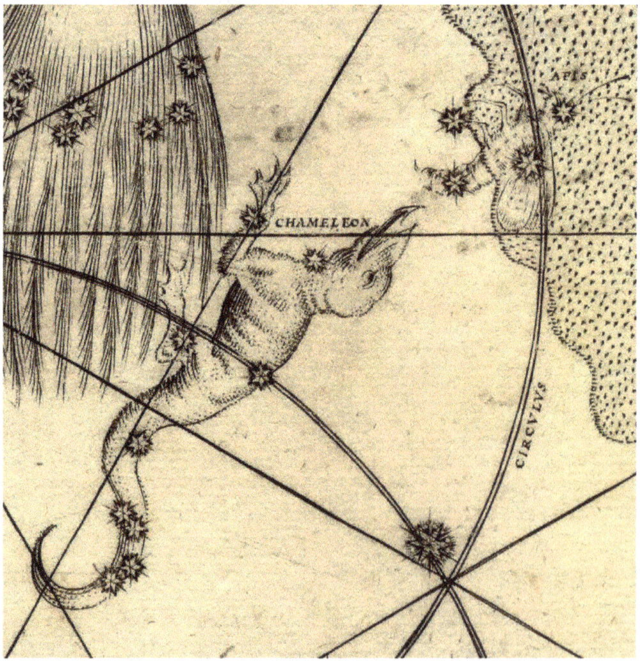

The constellation Musca (as Apis) is upper right. Detail from Johann Bayer's *Uranometria*, 1603.

In astronomy, constellations named after arthropods include the zodiacal Scorpius, the scorpion,and Musca, the fly, also known as Apis, the bee, in the deep southern sky. Musca, the only recognised insect constellation, was named by Petrus Plancius in 1597.

The Butterfly Nebula, NGC 6302.

"The Bug Nebula", also called "The Butterfly Nebula", is a more recent discovery. Known as NGC 6302 is one of the brightest and most popular stars in the universe – popular in that its features draw the attention of a lot of researchers. It happens to be located in the Scorpius constellation. It is perfectly bipolar, and until recently, the central star was unobservable, clouded by gas, but estimated to be one of the hottest in the galaxy – 200,000 degrees Fahrenheit, perhaps 35 times hotter than our Sun.

The honey bee played a central role in the cosmology of the Mayan people. The stucco figure at the temples of Tulum known as Ah Mucen Kab – the Diving Bee God – bears resemblance to the insect in the Codex Tro-Cortesianus identified as a bee. Such reliefs might have indicated towns and villages that produce honey. Modern Mayan authorities say the figure also have a connection to modern cosmology. Mayan mythology expert Migel Angel Vergara relates that the Mayans held a belief that bees came from Venus, the Second Sun. The relief might be indicative of another insect deity, that of Xux Ex, the Mayan wasp star. The Mayan embodied Venus in the form of the god Kukulkán (also known as or related to Gukumatz and Quetzalcoatl in other parts of Mexico), Quetzalcoatl is a Mesoamerican deity whose name in Nahuatl means feathered serpent. The cult was the first Mesoamerican religion to transcend the old Classic Period linguistic and ethnic divisions. This cult facilitated communication and peaceful trade among peoples of many different social and ethnic backgrounds. Although the cult was originally centered on the ancient city of Chichén Itzá in the modern Mexican state of Yucatán, it spread as far as the Guatemalan highlands.

In the Koran, the honeybee is the only creature that speak directly to God. Mohammed wrote in the 68–69 verses:

And your Lord taught the honey bee to build its cells in hills, on trees, and in (men's) habitations; Then to eat of all the produce (of the earth), and find with skill the spacious paths of its Lord: there issues from within their bodies a drink of varying colours, wherein is healing for men: verily in this is a Sign for those who give thought. (Surat an-Nahl (The Bee), 68–69)

In Costumes

A bee costume for a Mardi Gras celebration.

Bee and other insect costumes are worn in a variety of countries for parties, carnivals and other celebrations.

Ovo is an insect-themed production by the world renowned Canadian entertainment company Cirque du Soleil. The show looks at the world of insects and its biodiversity where they go about their daily lives until a mysterious egg appears in their midst, as the insects become awestruck about this iconic object that represents the enigma and cycles of their lives. The costuming was a fusion of arthropod body types blended with superhero armour. Liz Vandal, the lead costume designer, has a special affinity for the world of the insect:

When I was just a kid I put rocks down around the yard near the fruit trees and I lifted them regularly to watch the insects who had taken up residence underneath them. I petted caterpillars and let butterflies into the house. So when I learned that OVO was inspired by insects, I immediately knew that I was in a perfect position to pay tribute to this majestic world with my costumes. All insects are beautiful and perfect; it is what they evoke for each of us that changes our perception of them.

The Webby award-winning video series *Green Porno* was created to showcase the reproductive habits of insects. Jody Shapiro and Rick Gilbert were responsible for translating the research and concepts that Isabella Rossellini envisioned into the paper and paste costumes which directly contribute to the series' unique visual style. The film series was driven by the creation of costumes to translate scientific research into "something visual and how to make it comical."

REFERENCES

1. Posey, Darrell Addison (1986). "Topics And Issues In Ethnoentomology With Some Suggestions For The Development Of Hypothesis-Generation And Testing In Ethnobiology" (PDF). Journal of Ethnobiology. **6** *(1): 99–120. Archived (PDF) from the original on 1 August 2014.*
2. Schimitschek, E. (1968). Insekten als Nahrung, Brauchtum, Kult und Kultur. In Helmcke J.G.; Stark D.; Wermuth H. Handbuch der Zoologie. **4**. Berliner Akademie Verlag. pp. 1–62.
3. Hogue, Charles. Cultural Entomology. Retrieved 22 July 2012.
4. Who? What? Why?. Cultural Entomology Digest. Retrieved 22 July 2012.
5. Schimitschek, E. (1961). Die Bedeutung der Insekten für Kultur und Wirtschaft des Menschen in Vergangenheit und Gegenwart. Istanbul Universitesi. **908:**: 1–48.
6. Hogue, Charles (January 1987). Cultural Entomology. Annual Review of Entomology. **32**: 181–199. doi:10.1146/annurev.en.32.010187.001145.
7. Hogue, Charles. "Cultural Entomology". Cultural Entomology Digest. Retrieved 22 July 2012.
8. Morris, 2006.
9. Millennium Ecosystem Assessment (2005). Ecosystems and human well-being : synthesis (PDF). Washington, DC: Island Press. ISBN 1-59726-040-1. Retrieved 7 August 2014.
10. Lyman Kirst, Marian (29 Dec 2011). Insect – the neglected 99%. High Country News. Retrieved 22 July 2012.
11. Edward O. Wilson Biography. Biography.com. Retrieved 6 January 2017.
12. Wilson, E.O. (2006). The Creation: An Appeal to Save Life on Earth. Norton.
13. Little Creatures Who Run the World. Nova. PBS Airdate: August 12, 1997.
14. Higgins, Adrian (30 June 2007). Saving Earth From the Ground Up. June 30, 2007. The Washington Post. Retrieved 22 July 2012.

15. Kellert, S. R. (1993). Values and perceptions of invertebrates. Conservation Biology. 7 *(4): 845–855. doi:10.1046/j.1523-1739.1993.740845.x.*

16. Meyer, John (2006). Chapter 18: Insects As Pests. North Carolina State University. Retrieved 22 July 2012.

17. Miller G.T. (2004). Sustaining the Earth (6th ed.). Thompson Learning. pp. 211–216.

18. Meyer-Rochow, Victor Benno (1975). Can insects help to ease the problem of world food shortage?. *ANZAAS Journal: "Search". 6(7)*:: 261–262.

19. Chakravorty, J., Ghosh, S., and V.B. Meyer-Rochow. (2011). Practices of entomophagy and entomotherapy by members of the Nyishi and Galo tribes, two ethnic groups of the state of Arunachal Pradesh (North-East India). Journal of Ethnobiology and Ethnomedicine 7(5)

20. Meyer-Rochow V.B., Nonaka K., Boulidam S (2008) More feared than revered: Insects and their impacts on human societies (with specific data on the importance of entomophagy in a Laotian setting. Entomologie Heute 20: 3–25

21. Gullan, P. J.; Cranston, P. S. (2009). The Insects: An Outline of Entomology. John Wiley & Sons. pp. 9–13. ISBN 978-1-4051-4457-5.

22. Holland, Jennifer (14 May 2013). U.N. Urges Eating Insects: 8 Popular Bugs to Try. National Geographic. Retrieved 16 July 2015.

23. Haris, Emmaria (6 December 2013). Sensasi Rasa Unik Botok Lebah yang Menyengat (Unique taste sensation botok with stinging bees)(in Indonesian). Sayangi.com. Retrieved 22 June 2015.

24. Worms! A look at Zimbabwe's favorite snack: mopane worms. New York Daily News. 25 January 2013. Retrieved 10 July 2016.

25. Robinson, Martin; Bartlett, Ray; Whyte, Rob. Korea (2007). Lonely Planet publications, ISBN 1-74104-558-4, ISBN 978-1-74104-558-1. page 63

26. Tang, Philip. The 10 tastiest insects and bugs in Mexico. Lonely Planet. Retrieved 10 July 2016.

27. Isaacs, Jennifer (2002). Bush Food: Aboriginal Food and Herbal Medicine. Frenchs Forest, New South Wales: New Holland Publishers (Australia). pp. 190–192. ISBN 1-86436-816-0.

28. Bray, Adam (24 August 2010). Vietnam's most challenging foods. CNN: Travel. Retrieved 2 June 2015.

29. Kenyon, Chelsie. Chapulines. Retrieved 31 March 2015.

30. Fromme, Alison (2005). Edible Insects. Smithsonian Zoogoer. Smithsonian Institution. **34** *(4). Archived from the original on November 11, 2005. Retrieved 26 April 2015.*

31. Hanboonsong, Yupa; Tasanee, Jamjanya (2013-03-01). Six-legged livestock (PDF). FAO. Patrick B. Durst. Food and Agriculture Organization. Retrieved 10 July 2016.

32. Ramos-Elorduy de Concini, J. and J.M. Pino Moreno. (1988). The utilization of insects in the empirical medicine of ancient Mexicans. Journal of Ethnobiology, 8(2), 195–202.

33. http://tcm.health-info.org/Herbology.Materia.Medica/wugong-properties.htm

34. http://news.bbc.co.uk/2/hi/health/1809450.stm

35. Srivastava, S.K., Babu, N., and H. Pandey. (2009). Traditional insect bioprospecting—As human food and medicine. Indian Journal of Traditional Knowledge, 8(4): 485–494.

36. Feng, Y., Zhao, M., He, Z., Chen, Z., and L. Sun. (2009). Research and utilization of medicinal insects in China. Entomological Research, 39: 313–316.

37. Ratcliffe, N.A. et al. Insect Biochemistry and Molecular Biology, 41 (2011) 747e769

38. Dossey, A.T., 2010. Insects and their chemical weaponry: new potential for drug discovery. Nat. Prod. Rep 27, 1737e1757.

39. Sun, Xinjuan; Jiang, Kechun; Chen, Jingan; Wu, Liang; Lu, Hui; Wang, Aiping; Wang, Jianming (2014). A systematic review of maggot debridement therapy for chronically infected wounds and ulcers. International Journal of Infectious Diseases. **25**: 32–7. doi:10.1016/j.ijid.2014.03.1397. PMID 24841930.

40. Yang, X., Hu, K., Yan, G., et al., 2000. Fibrinogenolytic components in Tabanid, an ingredient in traditional Chinese medicine and their properties. J. Southwest Agric. Univ. 22, 173e176 (Chinese).

41. Ribeiro, J.M.C., Arca, B., 2009. From sialomes to the sialoverse: an insight into salivary potion of blood-feeding insects. Adv. Insect Physiol. 37, 59e118.

42. Francischetti, I.M.B., Mather, T.N., Ribeiro, J.M.C., 2005. Tick saliva is a potent inhibitor of endothelial cell proliferation and angiogenesis. Thromb. Haemost. 94, 167e174.

43. Maritz-Olivier, C., Stutzer, C., Jongejan, F., et al., 2007. Tick anti-hemostatics: targets for future vaccines and therapeutics. Trends Parasitol. 23, 397e407.

44. Carneiro-Lobo, T.C., Konig, S., Machado, D.E., 2009. Ixolaris, a tissue factor inhibitor, blocks primary tumor growth and angiogenesis in a glioblastoma model. J. Thromb. Haemost. 7, 1855e1864.

45. E. M. Costa-Neto, An. Acad. Bras. Cienc., 2005, 77(1), 33–43.

46. Pierce, BA (2006). Genetics: A Conceptual Approach (2nd ed.). New York: W.H. Freeman and Company. p. 87. ISBN 0-7167-8881-0.

47. Adams, MD; Celniker, SE; Holt, RA; Evans, CA; Gocayne, JD; Amanatides, PG; Scherer, SE; Li, PW; et al. (24 March 2000). The genome sequence of Drosophila melanogaster. Science. **287** *(5461): 2185–2195. Bibcode*:2000Sci...287.2185.. doi:10.1126/science.287.5461.2185. PMID 10731132.

48. Cochineal and Carmine. Major colourants and dyestuffs, mainly produced in horticultural systems. FAO. Retrieved June 16, 2015.

49. Guidance for Industry: Cochineal Extract and Carmine. FDA. Retrieved 6 July 2016.

50. How Shellac Is Manufactured. The Mail (Adelaide, SA : 1912 – 1954). 18 Dec 1937. Retrieved 17 July 2015.

51. Pearnchob, N.; Siepmann, J.; Bodmeier, R. (2003). Pharmaceutical applications of shellac: moisture-protective and taste-masking coatings and extended-release matrix tablets. Drug Development and Industrial Pharmacy. **29** *(8): 925–938.* doi:10.1081/ddc-120024188. PMID 14570313.

52. Barber, E. J. W. (1991). Prehistoric Textiles. Princeton University Press. pp. 230–231. ISBN 0-691-00224-X.

53. Goodwin, Jill (1982). A Dyer's Manual. Pelham. ISBN 0-7207-1327-7.

54. Schoeser, Mary (2007). Silk. Yale University Press. pp. 118, 121, 248. ISBN 0-300-11741-8.

55. Munro, John H. (2007). Netherton, Robin; Owen-Crocker, Gale R., eds. The Anti-Red Shift – To the Dark Side: Colour Changes in Flemish Luxury Woollens, 1300–1500. Medieval Clothing and Textiles. **3**. Boydell Press. pp. 56–57. ISBN 978-1-84383-291-1.

56. Munro, John H. (2003). Jenkins, David, ed. Medieval Woollens: Textiles, Technology, and Organisation. The Cambridge History of Western Textiles. Cambridge University Press. pp. 214–215. ISBN 0-521-34107-8.

57. Biomimetic Building Uses Termite Mound As Model. Retrieved 23 July 2012.

58. Termite research. esf.edu. Retrieved 23 July 2012.

59. Pappagallo, Linda. Termite Mounds Inspire Energy Neutral Buildings. Green Prophet. Retrieved 23 July 2012.

60. Namib Desert beetle inspires self-filling water bottle. BBC. 23 November 2012. Retrieved 14 June 2013.

61. Trivedi, Bijal P. Beetle's Shell Offers Clues to Harvesting Water in the Desert. National Geographic Today. Retrieved November 1, 2001.

62. Vainker, Shelagh (2004). Chinese Silk: A Cultural History. Rutgers University Press. p. 20. ISBN 0813534461.

63. Barber, E. J. W. (1992). Prehistoric textiles: the development of cloth in the Neolithic and Bronze Ages with special reference to the Aegean(reprint, illustrated ed.). Princeton University Press. p. 31. ISBN 978-0-691-00224-8. Retrieved 6 November 2010.

64. Department of Sericulture. History of Sericulture (PDF). Government of Andhra Pradesh (India). Archived from the original(PDF) on 21 July 2011. Retrieved 7 November 2010.

65. Bezzina, Neville. Silk Production Process. Sense of Nature Research. Archived from the original on 29 June 2012.

66. History of Silk. Silk road Foundation. Retrieved 7 January 2017.

67. Lockwood, Jeffrey A. (24 October 2008). Six-legged soldiers. The Scientist.

68. Lockwood, Jeffrey A. (2009). Six-Legged Soldiers: Using Insects as Weapons of War. Oxford University Press. ISBN 978-0-199-73353-8.

69. Lockwood, Jeffrey A. (21 October 2007). Bug Bomb. Why our next terrorist attack could come on six legs. The Boston Globe. Retrieved 6 January 2017.

70. Gullan, P.J.; Cranston, P.S. (2005). The Insects: An Outline of Entomology (3 ed.). Oxford: Blackwell Publishing. ISBN 1-4051-1113-5.

71. Shu, Catherine (27 February 2009). South Village welcomes spring with snacks — and an eye on environmental awareness. Taipei Times. Retrieved 7 January 2017.

72. Predicting Winter Weather: Woolly Bear Caterpillars. The Old Farmer's Almanac 2009. Retrieved 26 July 2012.

73. Frisch, Karl von (1956). Bees; their vision, chemical senses, and language. Cornell University Press.

74. Raffles, Hugh (2010). Insectopedia. Pantheon Books. p. 200.

75. Roth, L. M., and E. R. Willis. 1960. The biotic associations of cockroaches. Smithsonian Misc. Collect. 141:1–470.

76. What are Cockroaches?. Retrieved 21 August 2012.

77. What are the cockroach asthma antigens?. Curr Opin Pulm Med. 7: 14–9. Jan 2001. PMID 11140401.

78. Just Plain Weird Stories. Retrieved 21 August 2012.

79. Daddy Longleg myth. Myth Busters. Retrieved 21 August 2012.

80. Takada, Kenta (2012). Japanese Interest in Hotaru (Fireflies) and Kabuto-Mushi (Japanese Rhinoceros Beetles) Corresponds with Seasonality in Visible Abundance. Insects. **3** *(4): 423–431.* doi:10.3390/insects3020424.

81. William Balée (2000), "Antiquity of Traditional Ethnobiological Knowledge in Amazonia: a Tupí–Guaraní Family and Time Ethnohistory 47(2):399–422.

82. Kevin Groark. Taxonomic Identity of Hallucinogenic Harvester Ant (Pogonomyrmex californicus) Confirmed. 2001. Journal of Ethnobiology 21(2):133–144

83. Gough, Andrew. The Bee Part 2 Beewildered. Retrieved 23 July 2012.

84. A Beetle-Wing Tea-Cosy. Hampshire Cultural Trust. Retrieved 5 April 2016.

85. Rivers, Victoria. Beetles in Textiles. Cultural Entomology Digest February. Retrieved 23 July 2012.

86. Nabokov, Vladimir (1989). Speak, Memory. New York: Vintage International. p. 129.

87. Ahuja, Nitin. Nabokov's Case Against Natural Selection. Tract. Retrieved 23 July 2012.

88. Nabokov, Vladimir (1990). Strong Opinions. Vintage International.

89. MacRae, Ian. Butterfly Chronicles: Imagination and Desire in Natural & Literary Histories. Canadian Journal of Environ Education.

90. Kjell Sandved's Scaly Type. Retrieved 22 August 2012.
91. French, Anneka. Stirring the Swarm. Retrieved 21 August 2012.
92. Stirring the Swarm. Retrieved 21 August 2012.
93. Stagman, Myron (11 August 2010). Shakespeare's Greek Drama Secret. Cambridge Scholars Publishing. pp. 205–208. ISBN 978-1-4438-2466-8.
94. Walker, John Lewis (2002). Shakespeare and the Classical Tradition: An Annotated Bibliography, 1961–1991. Taylor & Francis. p. 363. ISBN 978-0-8240-6697-0.
95. Wells, H. G. (1904). Food of the Gods. Macmillan.
96. Rothman, Sheila; Rothman, David (2011). The Pursuit of Perfection: The Promise and Perils of Medical Enhancement. Knopf Doubleday. pp. 49–. ISBN 978-0-307-76713-4.
97. Hearn, Lafcadio *(2015). Insect Literature (Limited (300 copies)* ed.). Dublin: Swan River Press. pp. 1–18. ISBN 978-1-78380-009-4.
98. Nabokov, Vladimir (April 2000). Father's Butterflies. The Atlantic online. Retrieved 4 April 2016.
99. Amis, Martin (1994) [1993]. Visiting Mrs Nabokov: And Other Excursions. Penguin Books. pp. 115–18. ISBN 0-14-023858-1.
100. Boyd, Brian (April 2000). Nabokov's Butterflies, Introduction. The Atlantic. Retrieved 11 July 2016.
101. Gregersdotter, Katarina; Höglund, Johan; Hållén, Nicklas (2016). Animal Horror Cinema: Genre, History and Criticism. Springer. p. 147. ISBN 978-1-137-49639-3.
102. Warren, Bill; Thomas, Bill (2009). Keep Watching the Skies!: American Science Fiction Movies of the Fifties, The 21st Century Edition. McFarland. p. 32. ISBN 978-1-4766-2505-8.
103. Crouse, Richard (2008). Son of the 100 Best Movies You've Never Seen. ECW Press. p. 200. ISBN 978-1-55490-330-6.
104. All Things Considered (4 January 2005), 'Far Side' Entomology Class Has Students Abuzz, NPR, retrieved 9 July 2016
105. Larson, Gary (1989). The Prehistory of the Far Side. Andrews and McMeel. ISBN 0-8362-1851-5.
106. Rimsky-Korsakov – The Flight of the Bumblebee (Homeschool Music Appreciation). Retrieved 21 August 2012.
107. Eric Speed, Violinist, Flight of the Bumblebee, Nikon S9100, Montreal, 22 July 2011.. Retrieved 2012-04-21.
108. The Green Hornet. Retrieved 21 August 2012.
109. Neff,, Lyle (1999). The Tale of Tsar Saltan: A Centenary Appreciation of Rimski-Korsakov's Second Puškin Opera. The Pushkin Review. **2**: 89–133.
110. From the Diary of a Fly, for piano (Mikrokosmos Vol. 6/142), Sz.107/6/142, BB 105/142. Retrieved 21 August 2012.

111. David Rothenberg: Bug Music: How Insects Gave Us Rhythm and Noise. The Diana Rehm Show. Retrieved 2 April 2016.

112. Mark R. Chartrand III (1983) Skyguide: A Field Guide for Amateur Astronomers, p. 184 (ISBN 0-307-13667-1).

113. Constellations: A Guide to the Night Sky, http://www.constellation-guide.com/constellation-list/scorpius-constellation/

114. Ridpath, Ian. Frederick de Houtman's Catalogue. Star Tales. Retrieved 20 May 2016.

115. Matsuura, M.; Zijlstra, A. A.; Molster, F. J.; Waters, L. B. F. M.; Nomura, H.; Sahai, R.; Hoare, M. G. (2005). The dark lane of the planetary nebula NGC 6302. Monthly Notices of the Royal Astronomical Society. **359**: 383–400. Bibcode:2005MNRAS.359..383M. doi:10.1111/j.1365-2966.2005.08903.x.

116. Newly discovered star one of hottest in Galaxy. Jodrell Bank Centre for Astrophysics. Retrieved 26 July 2012.

117. Hunter, C. Bruce (1975). A Guide to Ancient Maya Ruins. University of Oklahoma Press. p. 309. ISBN 978-0-8061-1992-2.

118. Sharer, Robert J; Traxler, Loa P. (2006). The Ancient Maya (6th ed.). Stanford, CA: Stanford University Press. pp. 582–3. ISBN 0-8047-4817-9.

119. Sharer, Robert J; Traxler, Loa P. (2006). The Ancient Maya (6th ed.). Stanford, CA: Stanford University Press. p. 619. ISBN 0-8047-4817-9.

120. The Honey Bee. Retrieved 26 July 2012.

121. Philippines. Lonely Planet. 2000. p. 76. ISBN 978-0-86442-711-3.

122. Frank, Ze (2011). Young Me, Now Me: Identical Photos, Different Decades. Ulysses Press. pp. 23–. ISBN 978-1-61243-006-5.

123. About the Show. Cirque du Soleil. Retrieved 22 August 2012.

124. Kung, Michelle. Green Porno Creator Isabella Rossellini on Animal Sex, Crazed Fans and Web Entertainment. Wall Street Journal. Retrieved 2009-11-06.

CHAPTER 4

Forensic Entomology

Chapter Contents

- **Introduction**
- **History**
- **Forensic Entomology Subfields**
- **Invertebrate Types**
- **Factors**
- **Insect Activity Case Study**
- **In Literature**
- **Forensic Entomology and the Law**

Forensic Entomology is the use of the insects, and their arthropod relatives that inhabit decomposing remains to aid legal investigations. The broad field of forensic entomology is commonly broken down into three general areas: medicolegal, urban, and stored product pests. The medicolegal section focuses on the criminal component of the legal system and deals with the necrophilous (or carrion) feeding insects that typically infest human remains. The urban aspect deals with the insects that affect man and his immediate environment. This area has both criminal and civil components as urban pests may feed on both the living and the dead. The damage caused by their mandibles (or mouthparts) as they feed can produce markings and wounds on the skin that may be misinterpreted as prior abuse. Urban pests are of great economic importance and the forensic entomologist may become involved in civil proceedings over monetary damages. Lastly, stored product insects are commonly found in foodstuffs and the forensic entomologist may serve as an expert witness during both criminal and civil proceedings involving food contamination.

INTRODUCTION

Forensic entomology is the scientific study of the invasion and succession pattern of arthropods with their developmental stages of different species found on the decomposed cadavers during legal investigations. It is the application and study of insect and other arthropod biology to criminal matters. It also involves the application of the study of arthropods, including insects, arachnids, centipedes, millipedes, and crustaceans to criminal or legal cases. It is primarily associated with death investigations; however, it may also be used to detect drugs and poisons, determine the location of an incident, and find the presence and time of the infliction of wounds. Forensic entomology can be divided into three subfields: urban, stored-product and medico-legal/medico-criminal entomology.

HISTORY

Historically, there have been several accounts of applications for, and experimentation with, forensic entomology. The concept of forensic entomology dates back to at least the 13th century. However, only in the last 30 years has forensic entomology been systematically explored as a feasible source for evidence in criminal investigations. Through their own experiments and interest in arthropods and death, Sung Tzu, Francesco Redi, Bergeret d'Arbois, Jean Pierre Mégnin and the physiologist Hermann Reinhard have helped to lay the foundations for today's modern forensic entomology.

Sung Tzu

Sung Tzu (also known as Sung Tz'u) was a Judicial Intendant who lived in China 1188-1251 AD. In 1247 AD Sung Tzu wrote a book entitled *Washing Away of Wrongs* as a handbook for coroners. In this book Sung Tzu depicts several cases in which he took notes on how a person died and elaborates on probable causes. He explains in detail on how to examine a corpse both before and after burial. He also explains the process of how to determine a probable cause of death. The main purpose of this book was to be used as a guide for other investigators so they could assess the scene of the crime effectively. His level of detail in explaining what he observed in all his cases laid down the fundamentals for modern forensic entomologists and is the first recorded account in history of someone using forensic entomology for judicial means.

Francesco Redi

In 1668, Italian physician Francesco Redi disproved the theory of spontaneous generation. The accepted theory of Redi's day claimed that maggots developed spontaneously from rotting meat. In an experiment, he used samples of rotting meat that were either fully exposed to the air, partially exposed to the air, or not exposed to air at all. Redi showed that both fully and partially exposed rotting meat developed fly maggots, whereas rotting meat that was not exposed to air did not develop maggots. This discovery completely changed the way people viewed the decomposition of organisms and prompted further investigations into insect life cycles and into entomology in general.

Bergeret d'Arbois

Dr. Louis François Etienne Bergeret (1814–1893) was a French hospital physician, and was the first to apply forensic entomology to a case. In a case report published in 1855 he stated a general life cycle for insects and made many assumptions about their mating habits. Nevertheless, these assumptions led him to the first application of forensic entomology in an estimation of post-mortem interval (PMI). His report used forensic entomology as tool to prove his hypothesis on how and when the person had died.

Hermann Reinhard

The first systematic study in forensic entomology was conducted in 1881 by Hermann Reinhard, a German medical doctor who played a vital role in the history of forensic entomology. He exhumed many bodies and demonstrated that the development of many different types of insect species could be tied to buried bodies. Reinhard conducted his first study in east Germany, and collected many Phorid flies from this initial study. He also concluded that the development of only some of the insects living with corpses underground were associated with them, since there were 15-year-old beetles who had little direct contact with them. Reinhard's works and studies were used extensively in further forensic entomology studies.

Jean Pierre Mégnin

Jean Pierre Mégnin (1828–1905), an army veterinarian, published many articles and books on various subjects including the books *Faune des Tombeaux* and *La Faune des Cadavres*, which are considered to be among the most important forensic entomology books in history. In his second book he did revolutionary work on the theory of predictable waves, or successions of insects onto corpses. By counting numbers of live and dead mites that developed every 15 days and comparing this with his initial count on the infant, he was able to estimate how long that infant was dead.

In this book he asserted that exposed corpses were subject to eight successional waves, whereas buried corpses were only subject to two waves. Mégnin made many great discoveries that helped shed new light on many of the general characteristics of decaying flora and fauna. Mégnin's work and study of the larval and adult forms of insect families found in cadavers sparked the interest of future entomologists and encouraged more research in the link between arthropods and the deceased, and thereby helped to establish the scientific discipline of forensic entomology.

FORENSIC ENTOMOLOGY SUBFIELDS

Urban Forensic Entomology

Urban forensic entomology typically concerns pests infestations in buildings gardens or that may be the basis of litigation between private parties and service providers such as landlords or exterminators. Urban forensic entomology studies may also indicate the appropriateness of certain pesticide treatments and may also be used in stored products cases where it can help to determine chain of custody, when all points of possible infestation are examined in order to determine who is at fault.

Stored-product Forensic Entomology

Stored-product forensic entomology is often used in litigation over insect infestation or contamination of commercially distributed foods.

Medico-legal Forensic Entomology

Medicolegal forensic entomology covers evidence gathered through arthropod studies at the scenes of murder, suicide, rape, physical abuse and contraband trafficking. In murder investigations it deals with which insects eggs appear, their location on the body and in what order they appear. This can be helpful in determining a post mortem interval (PMI) and location of a death in question. Since many insects exhibit a degree of endemism (occurring only in certain places), or have a well-defined phenology (active only at a certain season, or time of day), their presence in association with other evidence can demonstrate potential links to times and locations where other events may have occurred. Another area covered by medicolegal forensic entomology is the relatively new field of entomotoxicology. This particular branch involves the utilization of entomological specimens found at a scene in order to test for different drugs that may have possibly played a role in the death of the victim.

INVERTEBRATE TYPES

Scorpionflies

Scorpionflies (order Mecoptera) were the first insects to arrive at a donated human cadaver observed (by the entomologist Natalie Lindgren) at the Southeast Texas Applied Forensic Science Facility near Huntsville, Texas, and remained on the corpse for one and a half days, outnumbering flies during that period. The presence of scorpionflies thus indicates that a body must be fresh.

Flies

Flies (order Diptera) are often first on the scene. They prefer a moist corpse for their offspring (maggots) to feed on. The most significant types of fly include:

- Blow flies – Family Calliphoridae- Flies in this family are often metallic in appearance and between 10 and 14 mm in length. In addition to the name blow-fly, some members of this family are known as blue bottle fly, cluster fly, greenbottles, or black blowfly. A characteristic of the blow-fly is its 3-segmented antennae. Hatching from an egg to the first larval stage takes from eight hours to one day. Larvae have three stages of development (called instars); each stage is separated by a molting event. Worldwide, there are 1100 known species of blowflies, with 228 species in the Neotropics, and a large number of species in Africa and Southern Europe. The most common area to find Calliphoridae species are in the countries of India, Japan, Central America, and in the southern United States. The typical habitat for blow-flies are temperate to tropical areas that provide a layer of loose, damp soil and litter where larvae may thrive and pupate. The forensic importance of this fly is that it is the first insect to come in contact with carrion because they have the ability to smell death from up to ten miles (16 km) away.

Flesh fly on decomposing flesh.

- Flesh flies – Family Sarcophagidae- Most flesh flies breed in carrion, dung, garbage, or decaying material, but a few species lay their eggs in the open wounds of mammals; hence their common name. Characteristics of the flesh-fly is its 3-segmented antennae. Most holarctic Sarcophagidae vary in size from 4 to 18 mm in length (Tropical species can be larger) with black and gray longitudinal stripes on the thorax and checkering on the abdomen. Flesh-flies, being viviparous, frequently give birth to live young on corpses of human and other animals, at any stage of decomposition, from newly dead through to bloated or decaying (though the latter is more common).
- House fly – Family Muscidae- is the most common of all flies found in homes, and indeed one of the most widely distributed insects; it is often considered a pest that can carry serious diseases. The adults are 6–9 mm long. Their thorax is gray, with four longitudinal dark lines on the back. The underside of their abdomen is yellow, and their whole body is covered with hair. Each female fly can lay up to 500 eggs in several batches of about 75 to 150 eggs. Genus *Hydrotaea* are of particular forensic importance.

- Cheese flies – Family Piophilidae - Most are scavengers in animal products and fungi. The best-known member of the family is *Piophila casei*. It is a small fly, about four mm (1/6 inch) long, found worldwide. This fly's larva infests cured meats, smoked fish, cheeses, and decaying animals and is sometimes called the cheese skipper for its leaping ability. Forensic entomology uses the presence of Piophila casei larvae to help estimate the date of death for human remains. They do not take up residence in a corpse until three to six months after death. The adult fly's body is black, blue-black, or bronze, with some yellow on the head, antennae, and legs. The wings are faintly iridescent and lie flat upon the fly's abdomen when at rest. At four mm (1/6 inch) long, the fly is one-third to one-half as long as the common housefly.
- Coffin flies – Phoridae
- Lesser corpse flies – Sphaeroceridae
- Lesser house flies – Fanniidae
- Black scavenger flies – Sepsidae
- Sun flies - Heleomyzidae
- Black soldier fly - Stratiomyidae - have potential for use in forensic entomology. The larvae are common scavengers in compost heaps, are found in association with carrion, can be destructive pests in honey bee hives, and are used in manure management (for both house fly control and reduction in manure volume). The larvae range in size from 1/8 to 3/4 of an inch (3 to 19 millimeters). The adult fly is a mimic, very close in size, color, and appearance to the organ pipe mud dauber wasp and its relatives.
- Phoridae – Humpbacked flies Larvae feed on decaying bodies. Some species can burrow to a depth of 50 cm over 4 days. Important in buried bodies.
- Non-biting midges - Chironomidae - these flies have a complex life cycle. While adults are terrestrial and phytophagous, larvae are aquatic and detritivorous. Immature instars have been used as forensic markers in several cases where submerged corpses were found.

Beetles

Beetles (Order Coleoptera) are generally found on the corpse when it is more decomposed. In drier conditions, the beetles can be replaced by moth flies (Psychodidae).

- Rove beetles – family Staphylinidae – are elongate beetles with small elytra (wing covers) and large jaws. Like other beetles inhabiting carrion, they have fast larval development with only three larval stages. Creophilus species are common predators of carrion, and since they are large, are a very visible component of the fauna of corpses. Some adult Staphylinidae are early visitors to a corpse, feeding on larvae of all species of fly, including the later predatory fly larvae. They lay their eggs in the corpse, and the emerging larvae are also predators. Some species have a long development time in the egg, and are common only during the later stages of decomposition. Staphylinids can also tear open the pupal cases of flies, to sustain themselves at a corpse for long periods.

- Hister beetles – family Histeridae. Adult histerids are usually shiny beetles (black or metallic-green) which have an introverted head. The carrion-feeding species only become active at night when they enter the maggot-infested part of the corpse to capture and devour their maggot prey. During daylight they hide under the corpse unless it is sufficiently decayed to enable them to hide inside it. They have fast larval development with only two larval stages. Among the first beetles to arrive at a corpse are Histeridae of the genus *Saprinus*. *Saprinus* adults feed on both the larvae and pupae of blowflies, although some have a preference for fresh pupae. The adults lay their eggs in the corpse, inhabiting it in the later stages of decay.
- Carrion beetles – family Silphidae- Adult Silphidae have an average size of about 12 mm. They are also referred to as burying beetles because they dig and bury small carcasses underground. Both parents tend to their young and exhibit communial breeding. The male carrion beetle's job in care is to provide protection for the breed and carcass from competitors.
- Ham beetles – family Cleridae
- Carcass beetles – family Trogidae
- Skin/hide beetles – family Dermestidae. Hide beetles are important in the final stages of decomposition of a carcass. The adults and larvae feed on the dried skin, tendons and bone left by fly larvae. Hide beetles are the only beetle with the enzymes necessary for breaking down keratin, a protein component of hair.
- Scarab beetles – family Scarabaeidae- Scarab beetles may be any one of around 30,000 beetle species worldwide that are compact, heavy-bodied and oval in shape. The flattened plates, which each antenna terminates, are fitted together to form a club. The outer edges of the front legs may also be toothed or scalloped. Scarab beetles range from 0.2 to 4.8 in (5.1 to 121.9 mm) in length. These species are known for being one of the heaviest insect species.
- Sap beetles – family Nitidulidae

Mites

Many mites (class Acari, not insects) feed on corpses with *Macrocheles* mites common in the early stages of decomposition, while Tyroglyphidae and Oribatidae mites such as Rostrozetes feed on dry skin in the later stages of decomposition.

Nicrophorus beetles often carry on their bodies the mite *Poecilochirus* which feed on fly eggs. If they arrive at the corpse before any fly eggs hatch into maggots, the first eggs are eaten and maggot development is delayed. This may lead to incorrect PMI estimates. *Nicrophorus* beetles find the ammonia excretions of blowfly maggots toxic, and the *Poecilochirus* mites, by keeping the maggot population low, allow *Nicrophorus* to occupy the corpse.

Moths

Moths (order Lepidoptera) specifically clothes-moths – Family Tineidae – are closely related to butterflies. Most species of moth are nocturnal, but there are crepuscular and diurnal species. Moths feed on mammalian hair during their larval stages and may forage on any hair that remains on a body. They are amongst the final animals contributing to the decomposition of a corpse.

Wasps, Ants, and Bees

Wasps, ants, and bees (order Hymenoptera) are not necessarily necrophagous. While some feed on the body, some are also predatory, and eat the insects feeding on the body. Bees and wasps have been seen feeding on the body during the early stages. This may cause problems for murder cases in which larval flies are used to estimate the post mortem interval since eggs and larvae on the body may have been consumed prior to the arrival on scene of investigators.

FACTORS

Moisture levels

Rain and humidity levels in the area where the body is found can affect the time for insect development. In most species, large amounts of rain will indirectly cause slower development due to drop in temperature. Light rain or a very humid environment, by acting as an insulator, will permit a greater core temperature within the maggot mass, resulting in faster development.

Submerged Corpses

M. Lee Goff, a noted and well respected forensic entomologist, was assigned to a case involving the discovery of a decomposing body found on a boat half a mile from shore. Upon collection of the maggot mass, only one insect, *Chrysomya megacephala*, was discovered. He concluded that the water barrier accounted for the scarcity of other flies. He also noted that flies will not attempt to trek across large bodies of water unless there is a substantially influential attractant.

In addition, the amount of time a maggot mass has been exposed to salt water can affect its development. From the cases Goff observed he found that if subjected for more than 30 minutes, there was a 24hour developmental delay. Unfortunately, not many more studies have been conducted and thus a specific amount of delay time is difficult to estimate.

Sun Exposure

"Because insects are cold-blooded animals, their rate of development is more or less dependent on ambient temperature." Bodies exposed to large amounts of sunlight will heat up, giving the insects a warmer area to develop, reducing their development time. An experiment conducted by Bernard Greenberg and John Charles Kunich with the use of rabbit carcasses to study accumulation of degree days found that with temperature ranging in the mid 70s to high 80s the amount of developmental time for maggots was significantly reduced.

In contrast, bodies found in shaded areas will be cooler, and insects will require longer growth periods. In addition, if temperatures reach extreme levels of cold, insects instinctively know to prolong their development time in order to hatch into a more accepting and viable climate in order to increase the chance of survival and reproduction.

Air Exposure

Hanged bodies can be expected to show their own quantity and variety of flies. Also, the amount of time flies will stay on a hanged body will vary in comparison to one found on the ground. A hanged body is more exposed to air and thus will dry out faster leaving less food source for the maggots.

As the body begins to decompose, a compilation of fluids will leak to the ground. In this area most of the expected fauna can be found. Also, it is more likely that rove beetles and other non-flying insects will be found here instead of directly on the body. Fly maggots, initially deposited on the body, may also be found below.

Geography

According to Jean Pierre Mégnin's book *La Faune des Cadavres* there are eight distinct faunal successions attracted to a corpse. While most beetles and flies of forensic importance can be found worldwide, a portion of them are limited to a specific range of habitats. It is forensically important to know the geographical distribution of these insects is order to determine information such as post mortem interval or whether a body has been moved from its original place of death.

Calliphoridae is arguably the most important family concerning forensic entomology given that they are the first to arrive on the corpse. The family's habitat ranges into the southern portion of the United States. However, while *Chrysomya rufifaces*, the hairy maggot blow fly, is part of the Calliphoridae family and is widespread, it is not prevalent in the Southern California, Arizona, New Mexico, Louisiana, Florida, or Illinois regions.

Flesh flies fall under the family Sacrophagidae and generally arrive to a corpse following Calliphoridae. However, as previously mentioned they are capable of flying in the rain. This key advantage enables them to occasionally reach a body before Calliphoridae overall effecting the maggot mass that will be discovered. Flesh flies are globally distributed including habitats in the United States, Europe, Asia, and the Middle East.

Beetles are representative of the order Coleoptera which accounts for the largest of the insect orders. Beetles are very adaptive and can be found in almost all environments with the exception of Antarctica and high mountainous regions. The most diverse beetle fauna can be found in the tropics. In addition, beetles are less submissive to temperatures.

Thus, if a carcass has been found in cold temperatures, the beetle will be prevalent over Calliphoridae.

Weather

Various weather conditions in a given amount of time cause certain pests to invade human households. This is because the insects are in search of food, water, and shelter. Damp weather causes reproduction and growth enhancement in many insect types, especially when coupled with warm temperatures. Most pests concerned at this time are ants, spiders, crickets, cockroaches, ladybugs, yellowjackets, hornets, mice, and rats. When conditions are dry, the deprivation of moisture outside drives many pests inside searching for water. While the rainy weather increases the numbers of insects, this dry weather causes pest invasions to increase. The pests most commonly known during dry conditions are scorpions, ants, pillbugs, millipedes, crickets, and spiders. Extreme drought does kill many populations of insects, but also drives surviving insects to invade more often. Cold temperatures outside will cause invasions beginning in the late summer months and early fall. Box elder bugs, cluster flies, ladybugs, and silverfish are noticed some of the most common insects to seek the warm indoors.

Modern techniques

Many new techniques have been developed and are used in order to more accurately gather evidence, or reevaluate at old information. The use of these newly developed techniques and evaluations have become relevant in litigation and appeals. Forensic entomology not only uses arthropod biology, but it pulls from other sciences, introducing fields like chemistry and genetics, exploiting their inherent synergy through the use of DNA in forensic entomology.

Scanning Electron Microscopy

Fly larvae and fly eggs are used to aid in the determination of a PMI. In order for the data to be useful the larvae and eggs must be identified down to a species level to get an accurate estimate for the PMI. There are many techniques currently being developed to differentiate between the various species of forensically important insects. A study in 2007 demonstrates a technique that can use scanning electron microscopy (SEM) to identify key morphological features of eggs and maggots. Some of the morphological differences that can help identify the different species are the presence/absence of anastomosis, the presence/absence of holes, and the shape and length of the median area.

The SEM method provides an array of morphological features for use in identifying fly eggs; however, this method does have some disadvantages. The main disadvantage is that it requires expensive equipment and can take time to identify the species from which the egg originated, so it may not be useful in a field study or to quickly identify a particular egg. The SEM method is effective provided there is ample time and the proper equipment and the particular fly eggs are plentiful. The ability to use these morphological differences gives forensic entomologists a powerful tool that can help with estimating a post mortem interval, along with other relevant information, such as whether the body has been disturbed post mortem.

Potassium Permanganate Staining

When scanning electron microscopy is not available, a faster, lower cost technique is potassium permanganate staining. The collected eggs are rinsed with a normal saline solution and placed in a glass petri dish. The eggs are soaked in a 1% potassium permanganate solution for one minute and then dehydrated and mounted onto a slide for observation. These slides can be used with any light microscope with a calibrated eyepiece to compare various morphological features. The most important and useful features for identifying eggs are the size, length, and width of the plastron, as well as the morphology of the plastron in the area around the micropyle. The various measurements and observations when compared to standards for forensically important species are used to determine the species of the egg.

Mitochondrial DNA

In 2001, a method was devised by Jeffrey Wells and Felix Sperling to use mitochondrial DNA to differentiate between different species of the subfamily Chrysomyinae. This is particularly useful when working to determine the identity of specimens that do not have distinctive morphological characteristics at certain life stages.

Mock crime Scenes

A valuable tool that is becoming very common in the training of forensic entomologists is the use of mock crime scenes using pig carcasses. The pig carcass represents a human body and can be used to illustrate various environmental effects on both arthropod succession and the estimate of the post mortem interval.

Gene Expression Studies

Although physical characteristics and sizes at various instars have been used to estimate fly age, a more recent study has been conducted to determine the age of an egg based on the expression of particular genes. This is particularly useful in determining developmental stages that are not evidenced by change in size; such as the egg or pupa and where only a general time interval can be estimated based on the duration of the particular developmental stage. This is done by breaking the stages down into smaller units separated by predictable changed in gene expression. Three genes were measured in an experiment with *Drosophila melanogaster*: bicoid (bcd), slalom (sll), and chitin synthase (cs). These three genes were used because they are likely to be in varied levels during different times of the egg development process. These genes all share a linear relationship in regards to age of the egg; that is, the older the egg is the more of the particular gene is expressed. However, all of the genes are expressed in varying amounts. Different genes on different loci would need to be selected for another fly species. The genes expressions are mapped in a control sample to formulate a developmental chart of the gene expression at certain time intervals. This chart can then be compared to the measured values of gene expression to accurately predict the age of an egg to within two hours with a high confidence level. Even though this technique can be used to estimate the age of an egg, the feasibility and legal acceptance of this must be considered for it to be a widely utilized forensic technique. One benefit of this would be that it is like other DNA-based techniques so most labs would be equipped to conduct similar experiments without requiring new capital investment. This style of age determination is in the process of being used to more accurately find the age of the instars and pupa; however, it is much more complicated, as there are more genes being expressed during these stages. The hope is that with this and other similar techniques a more accurate PMI can be obtained.

INSECT ACTIVITY CASE STUDY

A preliminary investigation of insect colonization and succession on remains in New Zealand revealed the following results on decay and insect colonization.

Open field Habitat

This environment had a daily average maximum temperature of 19.4 °C (66.9 °F) and a daily minimum temperature of 11.1 °C (52.0 °F). The average rainfall for the first 3 weeks in this environment was 3.0 mm/day. Around days 17–45, the body began to start active decay. During this stage, the insect successions started with *Calliphora stygia*, which lasted until day 27. The larvae of *Chrysomya rufifacies* were present between the day 13 and day 47. The *H. rostrata*, larvae of *Lucilia sericata*, Psychodidae family, and sylvicola were found to occur relatively late in the body's decay.

Coastal Sand-dune Habitat

This environment had an average daily maximum temperature of 21.4 °C (70.5 °F) and minimum of 13.5 °C (56.3 °F). The daily average rainfall was recorded as 1.4 mm/day for the first 3 weeks. The post-decay time interval, beginning at day six after death and ending around day 15 after death, is greatly reduced from the average post-decay time, due to the high average temperature of this environment. Insects obtained late in the post-active stage include the *Calliphora quadrimaculata*, adult Sphaeroceridae, Psychodidae and Piophilidae (no larvae from this last family were obtained in recovery).

Native Bush Habitat

This environment had recorded daily average maximum and minimum temperatures were 18.0 °C (64.4 °F) and 13.0 °C (55.4 °F), respectively. The average rainfall in this habitat was recorded at 0.4 mm/day. After the bloat stage, which lasted until day seven after death, post-active decay began around day 14. In this habitat, the *H. rostrata*, adult Phoridae, Sylvicola larvae and adult were the predominant species remaining on the body during the pre-skeletonization stages.

IN LITERATURE

Throughout its history the study of forensic entomology has not remained an esoteric science reserved only for entomologists and forensic scientists. Early twentieth-century popular scientific literature began to pique a broader interest in entomology. The very popular ten-volume book series, Alfred Brehem's *Thierleben* (Life of Animals, 1876–1879) expounded on many zoological topics, including arthropods. The accessible writing style of French entomologist Jean-Henri Fabre was also instrumental in the popularization of entomology. His collection of writings *Souvenirs Entomologique*, written during the last half of the 19th century, is especially useful because of the meticulous attention to detail to the observed insects' behaviors and life cycles.

The real impetus behind the modern cultural fascination with solving crime using entomological evidence can be traced back to the works *Faune des Tombeaux* (Fauna of the Tombs, 1887) and *Les Faunes des Cadavres* (Fauna of Corpses, 1894) by French veterinarian and entomologistJean Pierre Mégnin. These works made the concept of the process of insect ecological succession on a corpse understandable and interesting to an ordinary reader in a way that no other previous scientific work had done. It was after the publication of Mégnin's work that the studies of forensic science and entomology became an established part of Western popular culture, which in turn inspired other scientists to continue and expand upon his research.

FORENSIC ENTOMOLOGY AND THE LAW

Forensic entomology deals with the collection of arthropodic evidence and its application, and through a series of tests and previously set of rules, general admissibility of said evidence is determined. Ultimately, the admissibility of forensic evidence is left up to the judgement of the court. To provide a strong basis of admissibility for the evidence, accurate documentation is essential so that there is no room for speculation as to the authenticity of the evidence. Given that admissibility is granted, expert witnesses may be called to a courtroom to either support or refute the conclusions that are derived from the evidence submitted.

Scientific Evidence

In order to determine what is and is not scientific evidence, a set of rules or parameters must be established for qualification. There are two basic principles that govern the acceptability of evidence.

Frye Test

One of these qualifying tests is the Frye "General Acceptability" test. The Frye Standard is the test encountered more commonly in most state proceedings. The Frye Standard states that the scientific community must only accept the evidence after it is supported by certain amount of valid information. This test is used frequently to disqualify sketchy information such as bite mark comparisons, due to the scientific community's general belief that this can be a subjective and unreliable source and could lead to false conviction or acquittal if used in court proceedings.

Daubert Test

The second test is the Daubert relevancy and reliability requirement. This federal test is applied to testimony given by expert witnesses. Despite the Daubert Standard's reliability, it is used to a lesser extent in state courts. The information provided by the witness must pass a two-part examination, it must be deemed relevant to the case at hand, and the witness must prove that it came from a reliable source. Evidence presented in court proceedings is subject to extreme scrutiny, so it is very important that the information presented to the judge and jury has been obtained using the Scientific Method. Although these two decisions have been used as standards for many years, much of the decision about what evidence makes it into the courtroom depends upon the decisions by the presiding judge.

Admissibility of Evidence and Judicial Prudence

Information is usually admitted as evidence if it helps to clarify at least one of the five components of analyzing physical evidence: the reconstruction of events or places, identification as used in scientific tests, recognition by separating relevant and irrelevant information, by individualization by proving it to be unique to a suspect or a victim, or usage of comparison analysis methods (DNA, fingerprints, etc.). All of these categories, though important, have a lot of gray area, allowing for the judge to have the final say on what a jury can see and hear in regards to a case.

Aversion to Scientific Evidence

Judges tend to use more caution when admitting scientific data than when considering other types of physical evidence. This has been linked to a concern that scientific data is given priority over other forms of information. Jurors have been known to take scientific evidence as fact, so judges must decide how reliable the information is before allowing a jury to take that information into account. Allowing too much scientific data makes a case less subjective and gives jurors less of a choice in their deliberation; therefore, some judges try to limit charts, graphs, and other forms of overwhelming scientific evidence in relation to a crime. This is one of the main reasons that voiceprints and polygraphs have been denied admissibility in the courtroom. They are both subjective components of scientific evidence, but presented as fact by prosecutors, the jurors may become misinformed on how reliable the data is, and therefore most judges exclude it. These techniques are still used heavily in detective work, even though they are not admissible in court. These techniques help officers to find other types of evidence that can be brought to court and can help to ensure themselves that they are following the correct suspect.

Other Approaches to Scientific Evidence that May Allow its Admittance

One way that some judges have attempted to admit scientific evidence is in its raw form. By allowing the jury to draw their own conclusions, without confusing jargon and charts, some judges are more amicable to this approach. This requires a basic level of knowledge on how to read the data, which can be supplemented with an expert witness who may offer their own opinion. This is a less harsh approach to admitting scientific evidence.

Judges also place a large weight on the amount of corroborating evidence that can be submitted to support each piece. Typically, a judge will not admit a piece of evidence unless it can be supported by expert testimony, witnesses or other physical evidence. One piece of evidence in a crime is somewhat meaningless. Many pieces must fit together and support the prosecution's (or the defense's) theory. Similarly, a judge will not grant admissibility to evidence that proves guilt or innocence without a doubt. This rule is eagerly applied to matters of scientific evidence because juries tend to place too much emphasis on certain evidence, particularly DNA evidence, which is believed to be fool-proof. It is preferred that the circumstances are reconstructed and confirmed in small pieces. The admission of these pieces is viewed as unfair to the other counsel and is denied. The goal of the court is to leave a decision to be made by the jury after all of the evidence has been presented. Large, sweeping evidentiary discoveries take the subjectivity of the evidence out, especially in regards to scientific discoveries.

Forensic Entomology as Evidence in Court

Using forensic entomology as evidence in criminal proceedings requires the examination of a few other judicial requirements.

Federal Rules of Evidence 702 and 703

Most entomological evidence, or evidence dealing with the presence of insect larvae or adults being found at a scene, enters through expert witness testimony. Because there is little common knowledge associated with the discovery of particular larvae at a crime, it is useful for an entomologist to come in and explain what they found and what that could mean in regards to the location of death, the time of death, and other various parts of the case. This type of witness testimony enters under The Federal Rules of Evidence 702 and 703. (Daubert v. Merrell Dow Pharmaceuticals)

Use of Repulsive and Inflammatory Evidence

One rule that applies very specifically to forensic evidence is that the courts ban inflammatory or repulsive information or photographs. This can be especially difficult for entomologists, because photographs of insects feeding on dead bodies tend to be a little controversial, but they can be crucial in making their part of the case clear. Extreme caution must be used when choosing any photographs of entomological evidence to share with a jury. This rule was put into place in order to avoid outraged jurors. Emotional responses to this type of evidence has proven to distract jurors from the issue at hand, possibly leading them to make decisions based on irrelevant data.

Other Important Considerations

Another important factor in creating admissible entomological evidence is creating a clear and consistent chain of custody. This means that any samples taken from a body or the scene of the crime in question must have been labeled correctly and the identity of those persons who had access to it must be recorded. This record should show the location and custody of the samples at all times – from collection to analysis. Typically, an entomologist on the scene will collect their samples and lock them in their scene kit as a way of making sure that no one else has access to them to tamper with the evidence, ensuring that the information gathered in the field is viable in court. If an accurate record of custody is not kept, then the presentation of this evidence could provide a reason for appeal.

Case Records and Reports

As is the case when collecting and handling evidence, documentation of the source, transfer, and state of said entomological evidence is crucial to a criminal investigation. An accurate record facilitates the tracking of individual pieces of evidence even if the item has been modified after the initial collection; modification can result by means of natural biological processes, like metamorphosis, or the manipulation of the insects by the lab during analysis. Accurate documentation also helps preserve the chain of custody for the evidence collected; it accounts for a detailed record of who has had access to the evidence, and the exact times and locations of when the evidence was moved and for what reason. It basically provides a step-by-step ranging from the time of collection to the court hearings.

Forensic entomologist (FE) records relating to a death scene investigation can be divided into eight categories: initial contact notes, evidence submitted, Death Scene Case Study Form, autopsy report, local weather records, Specimen Disposition and Identification Record, chain of custody record and receipts, and finally, the Case Study Final Report.

Initial Contact Notes

Initial contact notes include certain preliminary information and the fee charged. The following are key pieces of preliminary information: the contact officer's name, address, and phone number; the death scene's exact location; the type of habitat in that area; the corpse's phase of decomposition; and a general description of the manner of death. When recording the information gathered, the estimated post mortem interval (PMI) made by others at the crime scene should be disregarded in order to guarantee an unbiased PMI from the entomologist. Since the entomology expertise is usually requested well after the recovery of the corpse, inquiry about the kind and amount of insect activity observed and whether the body had already been moved should also be recorded. Finally, the entomologist and the law enforcement organization should mutually agree on a clear fee that appropriately covers all the services to be provided by the FE.

Evidence Submitted

Because it is not always possible for the forensic entomologist to be present at a crime scene to collect the appropriate evidence, it is important that clear and concise instructions are left for the handling of live and preserved specimens to those processing the scene. In these cases where the FE is not present at the scene, the transport of live and preserved specimens is necessary. As a rule, if specimens are in transport longer than an hour, they have to be kept in a chilled insulated container with enough coolant to keep them relatively inactive and stop their development. All evidence should be handled with a reasonable amount of security; from videotapes and photographs to actual specimens recovered from the field. In all cases, the evidence should be secured with a seal that ensures absolutely no tampering.

Death Scene Case Study Form

Like the name suggests, this form is an itemized checklist for the different conditions present at the crime scene. The conditions recorded include temperature, habitats, the presence/absence of sunlight, and other details specific to the crime scene. More often than not, this form merely repeats previously recorded observations. This is because the primary purpose of the Death Scene Case Study Form is to itemize the different findings to ensure that nothing is overlooked.

Autopsy Report

The autopsy report is the responsibility of the coroner, not the forensic entomologist and must be requested. The most important things to note are the manner and cause of death, where the insects were found on the body, and the extent of decomposition of the body. Because the ME is one of the first people to have access to the body, he/she is usually the one to retrieve specimen samples from different orifices on the body. The specific location of the insects and the extent of their activity should be duly noted on the report. This information helps the entomologist determine an accurate PMI based on the development of the specimens found.

Local Weather Records

Accurate weather reports are elemental in a death scene investigation, especially one where any type of developmental insect are involved. Even within a relatively small area, the weather can vary, so it is important that the source of climate conditions is extremely reliable and accurate. For an ideal estimation as per the general maturation of the live specimens found on the body, a record of the ground and ambient temperatures should be taken for several days after the corpse is removed from the scene. To ensure the utmost accuracy it is best to collect weather data from a weather station located as close as possible to the actual crime scene.

Specimen Disposition and Identification Record

Most of the specimens collected from a corpse at a crime scene are maggots. Insects found in this immature stage of development are incredibly difficult to identify because all maggots have the same appearance when in this stage. It is because of this that most live specimens collected are reared into their adult stage where identification is increasingly easier.

It is in the specimen disposition and identification record that all developmental changes are recorded. Meticulous records of the changes that the specimen undergoes are duly noted and explained. In the same manner, when dissecting the immature insect, proper records must also be taken. Following dissection and proper documentation, all the specimen species are placed together in a labeled vial.

Chain of Custody Record and Receipts

Evidence is never in one place only; eventually, it has to be transferred among two or more parties, whether it be between entomologists, from the original site of the crime scene to an analytical lab, or between law enforcement departments. Documentation of all these transfers have to be properly annotated on the chain-of-custody records to ensure that the "chain of evidence" is never broken. The receipts serve as verification for the whereabouts of the evidence at all times.

Case study final report

The final and most important component of a crime scene's documentation is the Case Study Final Report. It is a compilation of all the relevant information gathered in the previously mentioned records and reports.

On it, there will be such information as the contact person's name, address, and phone number, a proposed scenario for the crime and, all relevant details on the crime scene, as well as the last time the deceased was seen alive; as you recall, this is all information found in the initial contact notes. Other relevant information is a list of all evidence collected, including specimens and all identification and rearing procedures performed, a PMI estimate, and a summarized conclusion based on all the findings. Copies of this report are submitted to the contact officer, the forensic entomologist, and as a courtesy, to the medical examiner. If throughout the course of the investigation, any manipulation of the specimens occurred (i.e. dissecting, rearing, then a copy of the specimen disposition and identification form should also be included in the report).

It is important that the FE remains completely objective as he/she documents his findings and compiles all significant evidence and data into this final report. At times, the evidence may allow for several different interpretations, and to remain objective, the FE must explore all these plausible scenarios.

Expert Witness

When a forensic entomologist is called to the stand, his/her expertise is only requested so that he/she may refute or support the Case Study Final Report. It is imperative that objectivity predominates and is not lost as the trial goes on. Although it may be used for referencing and corroboration of evidence, the Case Study Final Report is seldom admitted into court proceedings.

Preparation for Court

The Case Study Final Report is the main focus so that ideal preparation may take place.; this report does, after all, contain all the entomological findings and conclusions relevant to the case. The report should be detailed thoroughly, but should still allow room for speculation, because forensic entomology is still a relatively new science, and it would be a bit overzealous for the forensic entomologist to announce full certainty on the subject; said speculation should not exist when it comes to the evidence. Disclosure of this report to the opposing side before the trial is recommended so that the defense and/or prosecution have time to consult another expert witness to look over the original findings. Another thing is that proper credit should be given where credit is due; if any of the conclusions derived by the forensic entomologist were seconded by another professional, that person's name and qualifications should be presented as well.

It is the responsibility of the forensic entomologist to adequately instruct on the collection and handling of all specimen evidence. Even though, it is upon them to try to ensure efficient collections, they cannot always prevent inept actions; because of this, the FE can refuse improper analyses and tests performed on the evidence. It is important that the FE is diligent with the procedures and collection, so that when in court, the opposing side can't question the findings.

In Court

First of all, appearances in court merit professional attire. Upon arrival, the FE will wait his turn to be called to the stand in a waiting room. After being sworn in, the supporting counsel will ask for the FE's qualifications. It is vital that all qualifications, awards, published works, fellowships and grants, and all merit are mentioned because this information is what is going to determine whether the forensic entomologist is qualified enough to be an expert witness.

As the expert witness is being walked through the evidence, he/she has to keep in mind that most people have no thorough knowledge of entomology, so the responses to the questions being asked and the explanations about the evidence have to be kept clear and concise and in a simplistic manner that informs and educates, not only the jurors, but the judge as well. It is vital that the attorney has a thorough grasp on all the concepts that are being explored so that he/she knows which questions to ask to carry across his point. In the same manner, the responses to the questions being asked by the opposing side should be kept as precise as possible, so as not to given the opposing attorney any ammunition to discredit the findings in the cross-examination.

If any type of photographic evidence is involved, the expert witness has to request permission from the judge or attorney prior to the trial to make sure that the photographs that are to be presented aren't prejudiced and have the ability to sway the jury. Finally, it is important that the expert witness realizes that replying "I don't know" to a question is perfectly fine; it is preferable to say that you don't have the expertise in that particular area than to claim that you do and make yourself look like a liar. Before leaving the witness stand, proper notification will be given to the expert witness to see if he is dismissed or if he will be recalled later.

Case Studies

A good example of a case where forensic entomology was successful in determining a postmortem interval is one involving a young prostitute and a thirty-year-old member of the Army. A family member had informed the police of her disappearance four days before her body was found. She had last been seen in the presence of this military man on May 31. A large number of adult flies and maggots were noticed and gathered inside and near the wounds of the young girl. A number of these maggots were kept and grown to be adult flies (reared), and others were taken in their present state. Pictures were taken and enlarged showing the flies around the body when it was found. Using this evidence along with correlating weather conditions known to be present around that time, an entomologist was able to give a postmortem interval. The evidence showed that the flies had arrived on May 31, which indicated a time period of four days since she had died. This information showed the army man to be the last person to be with the girl. When presented with the evidence authorities were able to get a confession out of the man. The use of forensic entomology was definitely a key factor in this case.

Another case where forensic entomology was used successfully deals with a body being reported to be thrown down a well. When authorities searched for the well in southern Indiana, it was clear they found the correct one. Thousands of flies where flying over material blocking the well. The pungent smell and the fact that the flies could not access the body created this swarm. Given that vast number of wells in the area, the presence of the flies indicated the body's presence allowing it to be found in a timely manner. The use of entomology is not only important in determining when a person has died, but also simply locating a corpse.

Famous Cases

A handful of well publicized criminal cases helped bring forensic entomology into the limelight and popularized it in the modern era.

Paul Bernardo

In 1992, the Canadian press went into a frenzy concerning "The Ken and Barbie Murders." In June 1991, Paul Bernardo and Karla Homolka tortured, raped, and killed fourteen-year-old Leslie Mahaffy. The brutal assaults of Mahaffy were recorded on tape by the couple and helped convict them at trial.

Ten months later in April 1992, Bernardo and Homolka again raped, tortured, and killed another teenage victim, Kristen French over a period of several days, all captured on video tape.

Once captured and put on trial, forensic entomology began to play an important role. In Paul Bernardo's trial, renowned forensic entomologist Dr. Neal Haskell helped establish a timeline for one of the victim's murders. Haskell utilized his understanding of insect development cycles to narrow a fifteen-day time interval down to a period of four days for the victim's time of death.

Being one of Canada's most publicized criminal cases, Dr. Haskell and his testimony brought forth forensic entomology.

David Westerfield

In the United States, forensic entomology became well known through the trial of David Westerfield in 2002 for the abduction and murder of seven-year-old Danielle van Dam. In this San Diego case, the courtroom became a battle over conflicting reports from four forensic entomologists: David Faulkner, Neal Haskell, M. Lee Goff, and Robert D. Hall. A fifth scientist, William Rodriguez, a forensic anthropologist, also testified on this issue. Danielle had vanished from her bedroom the night of February 1–2; her partially decomposed body was found February 27. Haskell, Hall, and Faulker argued that forensic entomology established that the body of was dumped in mid-February, during which time Westerfield was already under police surveillance (since February 5), while Goff said the victim's body was colonized by larvae earlier than the other three entomologists estimated (his dates were February 9 to 14). His dates were earlier as he used new data from Canada which gives a significantly slower development for that fly species. Richard Merritt, a forensic entomologist and professor at Michigan State University, stated that because each scientist has his own method, this can lead to disagreement, but a lot of the disagreements involve a variation of one or two days, not over a week and a half as in the Westerfield case: "If it's that big a time, someone screwed up".

All the entomologists, including the prosecution's expert (Goff), gave dates that excluded the defendant by a significant margin - at least four days. Yet the jury rejected that evidence and found him guilty. The foreman explained that he felt "the field was too subjective to exonerate Westerfield".

The prosecutor, Jeff Dusek, was highly critical of the forensic entomology evidence; in fact he dismissed the field outright.

Other Important Cases

A case that shaped Canadian judicial history and brought forensic entomology into the headlines was the exoneration of Steven Truscott forty-eight years after being indicted for murder. Forensic entomology also played a role in the investigation of the 1993 raid on David Koresh and the Branch Davidians. Due to the publicity given to these cases, forensic entomology was able to seep further into popular culture.

In 2004, the first use of forensic entomology in Brazil demonstrated its importance and potential if executed properly. Twenty-six corpses were found in an indigenous region of Rondonia, Brazil. The bodies of coal miners who came in contact with a group of natives and were brutally clubbed and speared were found. The bodies were excavated and later transported to be autopsied. Arrowheads, spears, and clothing were collected but no larvae, pupae, or casings were recovered from the crime scene. Once the cadavers were autopsied, only fifty-seven larvae were present on the bodies, and unfortunately, a maximum post-mortem interval could not be established. In part because of the humidity and high temperature environment, the bodies decayed quickly. Since only *P. fulvinota,* a species which colonizes all stages of decay was found on the bodies, no timeline of succession could be established. Had more thorough work been done on the part of the investigators or the use of DNA in forensic entomology been employed, a more conclusive post mortem interval could have been established through the casings of stage-specific arthropods.

The *CSI* effect and the Law

From the contributions of movies and television series, forensic interest has increased and multiple branches spur into various fields. Entomology, an important area of research for the agriculture and veterinary sciences, also aids in crime scene investigation and forensic chronology. Coined after the explosively popular CBS series *CSI*, the CSI effect encompasses the overwhelming impact investigative scientific documentary shows have had on society.

Some notable American institutions such as Michigan State University, North Carolina State University and Texas A&M University have recently dedicated entire degree paths geared towards training skilled and knowledgeable forensic entomologists. West Virginia University's largest major is Forensic Science with over 400 undergraduates.

From 2000 to 2008, there has been a significant increase in the amount of attention paid by the public to forensic entomology as a science. With more attention given to arthropod development, accurate hypotheses can be formed. By following life cycles and tracking seasonal changes entomologists are able to establish the location and weather conditions of an area, which are both particularly useful to homicide detectives.However, with global climate change and increasing temperatures the formulas used to calculate post-mortem intervals may become skewed.Forensic entomologists like Gail Anderson, of Simon Fraser University in British Columbia, Canada, can rely on these formulas to provide certainty in their predictions. It is possible for them to estimate time of death to the exact day even if the person has been dead for a matter of weeks.

The lines blur when pseudoscience dramatized for viewers feeds misinformation and a false sense of security. Problems with jurors arise when they feel unsatisfied. The jury, wanting more conclusive evidence, scrutinizes minute details always expecting some kind of forensic "smoking gun". After watching a season of *House M.D.* or *CSI*, the jurors sometimes feel qualified to make expert decisions. Additionally the TV prejudices could leave some jurors wanting a dramatic experience, which can delay progress.

Even before the trial, the *CSI* effect can train criminals how to remove evidence and how to stage plausible scenarios. Conversely, victims can intentionally leave crucial traces behind. In one case, a victim who knew she was going to be transported, allowed insects to bite her in hopes of being found and traced back to her area.

Current Research and Entomological Findings

The vast majority of forensic entomological casework consists of establishing a very concrete timeline of the events involved in the crime scene. The new set of data provided by forensic entomological casework undoubtedly provides a solid case for either the defense or the prosecution. But as seen in a new series of experiments, the evidence may not be as firm as previously suspected. The use of DNA in forensic entomology is a growing area of research which will become more prevalent in cases as more research is done. In the first experiment, DNA was extracted from a series of immature blue-bodied blowflies.(blow-fly) The age range of the blowflies varied from first instar larvae up to the third day pupal stage.Extracting the target or victim's DNA from the gut of the larvae and pupae becomes inconclusive as you reach the third pupal stage. Therefore, as most samples collected from the crime scene can properly denote time scene death, there is a chance that the results may be skewed due to improper specimens.

Once the proper specimens are collected, a series of DNA sequencing tests can be initiated. Several methods of DNA sequencing exist to further analyze the gut material of the insect and eventually track down human microsatellite DNA. The sequencing methods currently utilized by forensic labs include STR or short tandem repeat profiling and HVR or the hyper variable regions of DNA. The current DNA analysis on forensic cases and forensic research indicates that STR profiling is generally successful, whereas HVR sequencing proved to a better choice in DNA testing, as its results had a higher success rate. In new research on the DNA analysis; "STR typing of human DNA from fly larvae fed on decomposing bodies" STR profiling created a complete profile in seven of the thirteen cases, and two cases displayed an incomplete set of alleles whereas four of the cases showed incomplete profiling. On the other hand, HVR sequencing was successful in all cases except one, So there is a slight margin of error when utilizing the STR profiling method of DNA typing in forensic analysis.

REFERENCES

1. Verma K, Paul R (2016). Lucilia sericata (Meigen) and Chrysomya megacephala (Fabricius) (Diptera: Calliphoridae) Development Rate and its Implications for Forensic Entomology. J Forensic Sci Med. 2 (3): 146–150. doi:10.4103/2349-5014.191466.
2. R.H. van Gulik (2004) [1956]. T'and-Yin-Pi-Shih: Parallel cases from under the pear-tree (reprint ed.). Gibson Press. p. 18. ISBN 0-88355-908-0.
3. S. Tz'u., B.E. Mc Knight 1981, The Washing Away Of Wrongs, Center for Chinese Studies The University of Michigan, Pages 1-34
4. A History of Microbiology. Historique.net. 30 April 2003. Retrieved 12 March 2008.

5. Benecke M. (2001). A brief history of forensic entomology. Forensic. Sci. Int. 120 (1–2): 2–14. doi:10.1016/S0379-0738(01)00409-1. PMID 11457602.
6. Klotzbach H, Krettek R, et al. (2004). The history of forensic entomology in German-speaking countries. Forensic. Sci. Int. 144 (2–3): 259–263. doi:10.1016/j.forsciint.2004.04.062. PMID 15364399.
7. ^ Jump up to:[a] [b] [c] Catts, E. P.; Goff, M. L. (January 1992). Forensic Entomology in Criminal Investigations. Annual Review of Entomology. 37: 253–272. doi:10.1146/annurev.en.37.010192.001345. PMID 1539937.
8. Bledsoe 2008, personal interview
9. Insects help convict killer. Retrieved 1 April 2008.
10. Rutsch, Poncie (22 January 2015). Finding Crime Clues In What Insects Had For Dinner. NPR. Retrieved 22 June 2015.
11. Southeast Texas Applied Forensic Science Facility. STAFS.
12. https://www.theguardian.com/science/2010/sep/23/flies-murder-natural-history-museum
13. González Medina A, Soriano Hernando Ó, Jiménez Ríos G (2015). The Use of the Developmental Rate of the Aquatic Midge Chironomus riparius (Diptera, Chironomidae) in the Assessment of the Postsubmersion Interval. J. Forensic. Sci. 60 (3): 822–826. doi:10.1111/1556-4029.12707.
14. Midgley JM, Richards CS, Villet MH, 2010. The utility of Coleoptera in forensic investigations. In: Amendt J, Campobasso CP, Goff ML, Grassberger M, eds. Current concepts in forensic entomology. Heidelberg: Springer, 57-68.
15. Scott, Michelle Pellissier (January 1998). The ecology and behavior of burying beetles. Annual Review of Entomology. 43: 595–618. doi:10.1146/annurev.ento.43.1.595. PMID 15012399.
16. unknown, www.encyclopedia.com/doc/1B1-377894.html
17. González Medina A, González Herrera L, Perotti MA, Jiménez Ríos G (2013). Occurrence of Poecilochirus austroasiaticus (Acari: Parasitidae) in forensic autopsies and its application on postmortem interval estimation. Exp.Appl. Acarol. 59 (3): 297–305. doi:10.1007/s10493-012-9606-1.
18. Forensic entomology: use of insects to help solve crimes. Uwa.edu.au. 20 March 2008.
19. Vilet MH, Richards CS, Midgley JM, 2010. Contemporary precision, bias and accuracy of minimum post-mortem intervals estimated using development of carrion-feeding insects. In: Amendt J, Campobasso CP, Goff ML, Grassberger M, eds. Current concepts in forensic entomology. Heidelberg: Springer, 109-137
20. Goff, M. L. A Fly for the Prosecution. Cambridge, Massachusetts: Harvard University Press, 2000.
21. Catts, E. P. and N. H. Haskell, eds. Entomology & Death: A Procedural Guide. Joyce's Print Shop, Inc. 1990. p5.
22. Greenberg, Bernard, and John C. Kunich. Entomology and the Law. United

Kingdom: Cambridge University Press, 2002.

23. Whitworth, Terry (2006). Keys to Genera and Species of Blow Flies of America North of Mexico. Proceedings of the Entomological Society of Washington. 108 (3): 710.

24. Pape, Thomas. Catalog of the Sarcophagidae of the World. Memoirs on Entomology. Gainesville, FL: Associated, 1996. 288-289.

25. Pest Control Tips: Pests and Weather. Terminix.com. 19 March 2008.

26. Villet MH, Amendt J, 2011. Advances in entomological methods for estimating time of death. In: Turk EE, ed. Forensic Pathology Reviews. Heidelberg: Humana Press, pp. 213-238

27. Mendonça, Paloma Martins. Identification of fly eggs using scanning electron microscopy for forensic investigations. 2008. Micron. 13 March 2008

28. Sukontason, Kom; Sukontason, Kabkaew L; Piangjai, Somsak; Boonchu, Noppawan; Kurahashi, Hiromu; Hope, Michelle; Olson, Jimmy K (2004). "Identification of forensically important fly eggs using a potassium permanganate staining technique". Micron. 35 (5): 391–395. doi:10.1016/j.micron.2003.12.004. ISSN 0968-4328. PMID 15006363.

29. Wells, D. and Sperling Felix A. H. "DNA-based identification of forensically important Chrysomyinae (Diptera: Calliphoridae)Forensic Science International Volume 120, Issues 1-215 August 2001 110-115 . 3 March 2008

30. Schoenly, Kenneth G. Recreating Death's Acre in the School Yard: Using Pig Carcasses as Model American Biology Teacher v68 n7 September 2006 402-410 . 3 March 2008

31. Tarone, Aaron M.; Jennings, Kimberley C.; Foran, David R. (November 2007). "Aging Blow Fly Eggs Using Gene Expression: A Feasibility Study". Journal of Forensic Sciences. 52 (6): 1350–1354. doi:10.1111/j.1556-4029.2007.00587.x. PMID 18093065. Retrieved 1 April 2008.

32. Eberhardt TL, Elliot DA (2008). A preliminary investigation of insect colonisation and succession on remains in New Zealand. Forensic Sci. Int. 176 (2–3): 217–223. doi:10.1016/j.forsciint.2007.09.010. PMID 17997065.

33. Benecke, M. (2001). A brief history of forensic entomology. Forensic Entomology International, 120, page 8.

34. "Jean Henri-Fabre." Scarab Workers World Directory. January 1998 - January 2007. University of Nebraska-Lincoln State Museum-Division of Entomology. 13 March 2008. http://www.unl.edu/museum/research/entomology/workers/JFabre.htm

35. Benecke, M. (2001). A brief history of forensic entomology. Forensic Entomology International, 120, page 5

CHAPTER 5

Medical Entomology

Chapter Contents

- Introduction
- Insects of Medical Importance
- Personal Pests
- The Housefly
- The Cockroach
- Biting Insects
- Major Insect-Born Disease
- Bed Bug
- Scabies

Medical entomology: The branch of science that deals with insects that cause disease or that serve as vectors of organisms that cause disease in humans. In medical entomology, for example, research may pertain to the interactions of the arboviruses and their hosts and involve physiology, genetics, biology, and ecology.

INTRODUCTION

The discipline of **medical entomology**, or **public health entomology**, and also *veterinary entomology* is focused upon insects and arthropods that impact human health. Veterinary entomology is included in this category, because many animal diseases can "jump species" and become a human health threat, for example, bovine encephalitis. Medical entomology also includes scientific research on the behavior, ecology, and epidemiology of arthropod disease vectors, and involves a tremendous outreach to the public, including local and state officials and other stake holders in the interest of public safety, finally in current situation related to one health approach mostly health policy makers recommends to widely applicability of medical entomology for disease control efficient and best fit on achieving development goal and to tackle the newly budding zoonotic diseases. Thoughtful to have and acquaint with best practice of Med. Entomologist to tackle the animal and public health issues together with controlling arthropods born diseases by having Medical Entomologists' the right hand for bringing the healthy world .

A U.S. Navy medical entomologist identifying insects

Aedes albopictus.

Medical Entomologists are employed by private and public universities, private industries, and federal, state, and local government agencies, including all three branches of the US military - who hire medical entomologists to protect the troops from infectious diseases that can be transmitted by arthropods. Historically, during wars, more people have died due to insect-transmitted diseases, than to all the battle injuries combined.

Medical entomologists are also hired by chemical companies - to help develop new pesticides which will effectively decrease insect pest populations while simultaneously protecting the health of the public.

Public health entomology has seen a huge surge in interest since 2005, due to the resurgence of the bed bug, *Cimex lectularius*.

INSECTS OF MEDICAL IMPORTANCE

Medical entomologists work in the public health arena, dealing with insects (and other arthropods) that parasitize people, bite, sting, and/or vector disease.

PERSONAL PESTS

Some personal pests of may vector pathogens:Lice, Fleas, Bedbugs, Ticks, Scabies mites

THE HOUSEFLY

The housefly is a very common and cosmopolitan species which transmits diseases to man. The organisms of both amoebic and bacillary dysenteries are picked up by flies from the faeces of infected people and transferred to clean food either on the fly's hairs or by the fly vomiting during feeding. Typhoid germs may be deposited on food with the fly's faeces. The house fly cause the spread of yaws germs by carrying them from a yaws ulcer to an ordinary sore. Houseflies also transmit poliomyelitis by carrying the virus from infected faeces to food or drink. Cholera and hepatitis are sometimes fly-borne. Other diseases carried by houseflies are Salmonella, tuberculosis, anthrax, and some forms of ophthalmia. They carry over 100 pathogens and transmit some parasitic worms. The flies in poorer and lower-hygiene areas usually carry more pathogens. Some strains have become immune to most common insecticides.

THE COCKROACH

Cockroaches carry disease-causing organisms (typically gastroenteritis) as they forage. Cockroach excrement and cast skins also contain a number of allergens causing responses such as, watery eyes,skin rashes, congestion of nasal passages and asthma.

BITING INSECTS

Mosquitoes, Biting Midges, Sandflies, Black flies, Horse Flies, Stable flies.

Pathogen infection transmitted by insect or other arthropod vectors.

Diseases carried by insects and other arthropod vectors affect more than 700 million people every year, and are considered the most sensitive to climatic and environment conditions.(WHO)

MAJOR INSECT-BORN DISEASE

- Dengue fever - Vectors: *Aedes aegypti* (main vector) *Aedes albopictus* (minor vector) threatens -50 million people are infected by dengue annually, 25,000 die. Threatens 2.5 billion people in more than 100 countries.
- Malaria - Vectors: *Anopheles* mosquitoes - 500 million become severely ill with malaria every year and more than 1 million die.
- Leishmaniasis - Vectors: species in the genus *Lutzomyia* in the New World and *Phlebotomus* in the Old World. Two million people infected.
- Bubonic plague - Principal vector: *Xenopsylla cheopis* At least 100 flea species can transmit plague. Re-emerging major threat several thousand human cases per year.High pathogenicity and rapid spread.
- Sleeping sickness - Vector: Tsetse fly, not all species. Sleeping sickness threatens millions of people in 36 countries of sub-Saharan Africa (WHO)
- Typhus - Vectors: mites, fleas and body lice 16 million cases a year, resulting in 600,000 deaths annually.
- Wuchereria bancrofti - most common vectors: the mosquito species: *Culex*, *Anopheles*, *Mansonia*, and *Aedes*; affects over 120 million people.

- Yellow Fever - Principal vectors: *Aedes simpsoni*, *A. africanus*, and *A. aegypti* in Africa, species in *Haemagogus* genus in South America, and species in *Sabethes* genus in France -200,000 estimated cases of yellow fever (with 30,000 deaths) per year.

Minor

- Ross River fever - Vector: Mosquitoes, main vectors *A. vigilax*, *Aedes camptorhynchus*, and *Culex annulirostris*
- Barmah Forest Virus - Vector: Known vectors *Culex annulirostris*, *Ocleratus vigilax* and *O. camptorhynchus* and *Culicoides marksi*
- Kunjin encephalitis (mosquitoes)
- Murray Valley encephalitis virus (MVEV) - Major mosquito vector: *Culex annulirostris*.
- Japanese encephalitis - Several mosquito vectors, the most important being *Culex tritaeniorhynchus*.
- West Nile virus - Vectors: vary according to geographical area; in the United States *Culex pipiens* (Eastern US), *Culex tarsalis* (Midwest and West), and *Culex quinquefasciatus* (Southeast) are the main vectors.
- Lyme disease - Vectors: several species of the genus *Ixodes*
- Alkhurma virus (KFDV) - Vector: tick
- Kyasanur forest disease - Vector: *Haemaphysalis spinigera*
- Brugia timori filariasis - Primary vector: *Anopheles barbirostris*
- Babesia - Vector *Ixodes* ticks.
- Carrion's disease - Vectors: sandflies of the genus *Lutzomyia*.
- Chagas disease - Vector: assassin bugs of the subfamily *Triatominae*. The major vectors are species in the genera *Triatoma*, *Rhodnius*, and *Panstrongylus*.
- Chikungunya - Vectors: *Aedes* mosquitoes
- Human ewingii ehrlichiosis - Vector: *Amblyomma americanum*
- Human granulocytic ehrlichiosis - Vector: *Ixodes scapularis*
- Rift Valley Fever (RVF) - Vectors: fleas in the genera *Aedes* and *Culex*
- Scrub typhus - Vector: Chigger
- Loa loa filariasis - Vector: *Chrysops* sp.

LOUSE

Louse (plural: **lice**) is the common name for members of the order Phthiraptera, which contains nearly 5,000 species of wingless insect. Lice are obligate parasites, living externally on warm-blooded hosts which include every species of bird and mammal, except for monotremes, pangolins, bats and cetaceans. Lice are vectors of diseases such as typhus.

Chewing lice live among the hairs or feathers of their host and feed on skin and debris, while sucking lice pierce the host's skin and feed on blood and other secretions. They usually spend their whole life on a single host, cementing their eggs, which are known as nits, to hairs or feathers. The eggs hatch into nymphs, which moult three times before becoming fully grown, a process that takes about four weeks.

Humans host three species of louse, the head louse, the body louse and the pubic louse. The body louse has the smallest genome of any known insect; it has been used as a model organism and has been the subject of much research.

Lice were ubiquitous in human society until at least the Middle Ages. They appear in folktales, songs such as *The Kilkenny Louse House*, and novels such as James Joyce's *Finnegans Wake*. They commonly feature in the psychiatric disorder delusional parasitosis. A louse was one of the early subjects of microscopy, appearing in Robert Hooke's 1667 book, *Micrographia*.

Classification

The order Phthiraptera is clearly a monophyletic grouping, united as the members are by a number of derived features including their parasitism on warm-blooded vertebrates and the combination of their metathoracicganglia with their abdominal ganglia to form a single ventral nerve junction. The order has traditionally been divided into two suborders, the sucking lice (Anoplura) and the chewing (Mallophaga); however, recent classifications suggest that the Mallophaga are paraphyletic and four suborders are now recognized:

- Anoplura: sucking lice, occurring on mammals exclusively
- Rhynchophthirina: parasites of elephants and warthogs
- Ischnocera: mostly avian chewing lice, however, one family parasitizes mammals
- Amblycera: a primitive suborder of chewing lice, widespread on birds, however, also live on South-American and Australian mammals

Nearly 5,000 species of louse have been identified, about 4,000 being parasitic on birds and 800 on mammals. Lice are present on every continent in all the habitats that their host animals and birds occupy. They are found even in the Antarctic, where penguins carry 15 species of lice (in the genera *Austrogonoides* and *Nesiotinus*).

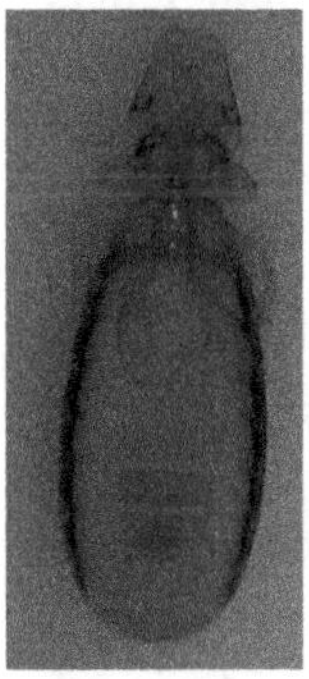

Ricinus bombycillae, an Amblyceran louse from the bohemian waxwing

Trinoton anserinum, an Amblyceran louse from a mute swan

Damalinia limbata is an Ischnoceran louse from goats. The species is sexually dimorphic, with the male smaller than the female.

Description

Sucking lice are small wingless insects ranging from 0.5 to 5 mm (0.02 to 0.20 in) in length. They have narrow heads and oval, flattened bodies. They have no ocelli, and their compound eyes are reduced in size or absent. Their antennae are short with three to five segments, and their mouth parts, which are retractable into their head, are adapted for piercing and sucking. There is a cibarial pump at the start of the gut; it is powered by muscles attached to the inside of the cuticle of the head. The mouthparts consist of a proboscis which is toothed, and a set of stylets arranged in a cylinder inside the proboscis, contaiing a salivary canal (ventrally) and a food canal (dorsally). The thoracic segments are fused, the abdominal segments are separate, and there is a single large claw at the tip of each of the six legs.

Chewing lice are also flattened and can be slightly larger than sucking lice, ranging in length from 0.5 to 6 mm (0.02 to 0.24 in). They are similar to sucking lice in form but the head is wider than the thorax and all species have compound eyes. There are no ocelli and the mouthparts are adapted for chewing. The antennae have three to five segments and are slender in the suborder Ischnocera, but club-shaped in the suborder Amblycera. The legs are short and robust, and terminated by one or two claws. Many lice are specific to a single species of host and have co-evolved with it. They are usually cryptically coloured to match the fur or feathers of the host.

Biology

Lice are divided into two groups: sucking lice, which obtain their nourishment from feeding on the sebaceous secretions and body fluids of their host; and chewing lice, which are scavengers, feeding on skin, fragments of feathers or hair, and debris found on the host's body. Most are found on only specific types of animals, and, in some cases, on only a particular part of the body; some animals are known to host up to fifteen different species, although one to three is typical for mammals, and two to six for birds. For example, in humans, different species of louse inhabit the scalp and pubic hair. Lice generally cannot survive for long if removed from their host. Some species of chewing lice house symbiotic bacteria in bacteriocytes in their bodies. These may assist in digestion because if the insect is deprived of them, it will die. If their host dies, lice can opportunistically use phoresis to hitch a ride on a fly and attempt to find a new host.

A louse's color varies from pale beige to dark gray; however, if feeding on blood, it may become considerably darker. Female lice are usually more common than males, and some species are parthenogenetic, with young developing from unfertilized eggs. A louse's egg is commonly called a nit. Many lice attach their eggs to their hosts' hair with specialized saliva; the saliva/hair bond is very difficult to sever without specialized products. Lice inhabiting birds, however, may simply leave their eggs in parts of the body inaccessible to preening, such as the interior of feather shafts. Living louse eggs tend to be pale whitish, whereas dead louse eggs are yellower.

Lice are exopterygotes, being born as miniature versions of the adult, known as nymphs. The young moult three times before reaching the final adult form, usually within a month after hatching.

Ecology

The average number of lice per host tends to be higher in large-bodied bird species than in small ones. Lice have an aggregated distribution across bird individuals, i.e. most lice live on a few birds, while most birds are relatively free of lice. This pattern is more pronounced in territorial than in colonial—more social—bird species. Host organisms that dive under water to feed on aquatic prey harbor fewer taxa of lice. Bird taxa that are capable of exerting stronger antiparasitic defense—such as stronger T cell immune response or larger uropygial glands—harbor more taxa of Amblyceran lice than others.

Reductions in the size of host populations may cause a long-lasting reduction of louse taxonomic richness, for example, birds introduced into New Zealand host fewer species of lice there than in Europe. Louse sex ratios are more balanced in more social hosts and more female-biased in less social hosts, presumably due to the stronger isolation among louse subpopulations (living on separate birds) in the latter case. The extinction of a species results in the extinction of its host-specific lice. Host-switching is a random event that would seem very rarely likely to be successful, but speciation has occurred over evolutionary time-scales so it must be successfully accomplished sometimes.

Lice may reduce host life expectancy if the infestation is heavy, but most seem to have little effect on their host. The habit of dust bathing in domestic hens is probably an attempt by the birds to rid themselves of lice. Lice may transmit microbial diseases and helminth parasites, but most individuals spend their whole life cycle on a single host and are only able to transfer to a new host opportunistically. Ischnoceran lice may reduce the thermoregulation effect of the plumage; thus heavily infested birds lose more heat than others. Lice infestation is a disadvantage in the context of sexual rivalry.

The human body louse *Pediculus humanus humanus* had its genome sequenced in 2010, and at that time it had the smallest insect genome known. This louse can transmit certain diseases while the human head louse (*P. h. capitis*), to which it is closely related, cannot. With their simple life history and small genomes, the pair make ideal model organisms to study the molecular mechanisms behind the transmission of pathogens and vector competence. In 2015 there were 5 cases of louse-borne relapsing fever being transmitted to locals reported in Italy.

Interaction with Humans

Prehistory

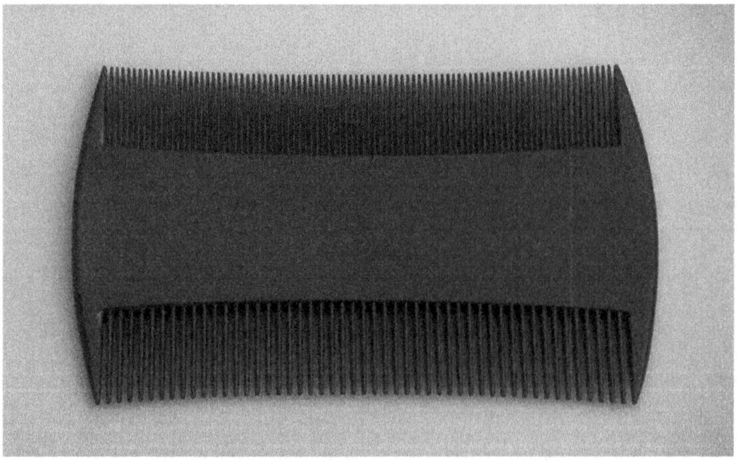

A louse (nit) comb.

Humans host three different kinds of lice: head lice, body lice, and pubic lice. Lice infestations can be controlled with lice combs, and medicated shampoos or washes.

Lice have been the subject of significant DNA research in the 2000s that led to discoveries on human evolution. The three species of sucking lice that parasitize human beings belong to two genera, *Pediculus* and *Phthirus*: head lice (*Pediculus humanus capitis*), body lice (*Pediculus humanus corporis*), and pubic lice (*Phthirus pubis*). Human head and body lice (genus *Pediculus*) share a common ancestor with chimpanzee lice, while pubic lice (genus *Phthirus*) share a common ancestor with gorilla lice. Using phylogenetic and cophylogenetic analysis, Reed et al. hypothesized that *Pediculus* and *Phthirus* are sister taxa and monophyletic. In other words, the two genera descended from the same common ancestor. The age of divergence between *Pediculus* and its common ancestor is estimated to be 6-7 million years ago, which matches the age predicted by chimpanzee-hominid divergence. Because parasites rely on their hosts, host-parasite cospeciation events are likely.

For example, genetic evidence suggests that our human ancestors acquired pubic lice from gorillas approximately 3-4 million years ago. Unlike the genus *Pediculus*, the divergence in *Phthirus* does not match the age of host divergence that likely occurred 7 million years ago. Reed et al. propose a *Phthirus* species host-switch around 3-4 million years ago. While it is difficult to determine if a parasite-host switch occurred in evolutionary history, this explanation is the most parsimonious (containing the fewest evolutionary changes).

Additionally, the DNA differences between head lice and body lice provide corroborating evidence that humans used clothing between 80,000 and 170,000 years ago, before leaving Africa. Human head and body lice occupy distinct ecological zones: head lice live and feed on the scalp, while body lice live on clothing and feed on the body. Because body lice require clothing to survive, the divergence of head and body lice from their common ancestor provides an estimate of the date of introduction of clothing in human evolutionary history.

The mitochondrial genome of the human species of body lice (*Pediculus humanus humanus*), the head louse (*Pediculus humanus capitis*) and the pubic louse (*Phthirus pubis*) fragmented into a number of minichromosomes, at least seven million years ago. Analysis of mitochondrial DNA in human body and hair lice reveals that greater genetic diversity existed in African than in non-African lice. Human lice can also shed light on human migratory patterns in pre-history. The dominating theory of anthropologists regarding human migration is the Out of Africa Hypothesis. Genetic diversity accumulates over time, and mutations occur at a relatively constant rate. Because there is more genetic diversity in African lice, the lice and their human hosts must have existed in Africa before anywhere else.

Modern History

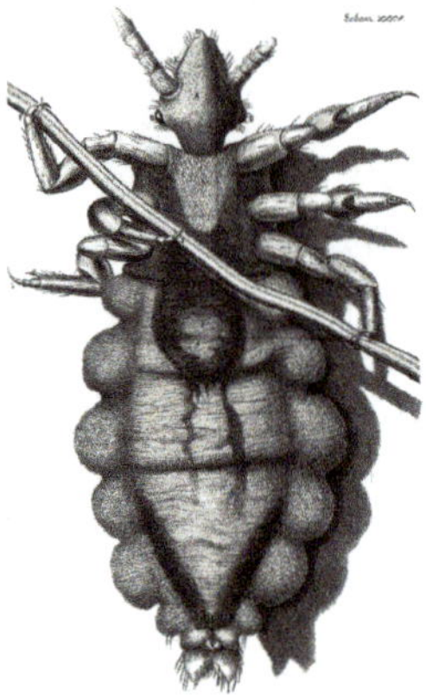

Drawing of a louse clinging to a human hair. Robert Hooke, *Micrographia*, 1667.

Lice have been intimately associated with human society throughout history. In the Middle Ages, they were essentially ubiquitous. At the death of Thomas Becket, Archbishop of Canterbury in 1270, it was recorded that "The vermin boiled over like water in a simmering cauldron, and the onlookers burst into alternate weeping and laughing". A mediaeval treatment for lice was an ointment made from pork grease, incense, lead, and aloe. Robert Hooke's 1667 book, *Micrographia: or some physiological descriptions of minute bodies made by magnifying glasses with observations and Inquiries thereupon*, illustrated a human louse, drawn as seen down an early microscope.

Margaret Cavendish's satirical *The Description of a New World, Called The Blazing-World* (1668) has "Lice-men" as "mathematicians", investigating nature by trying to weigh the air like the real scientist Robert Boyle.

In 1935 the Harvard medical researcher Hans Zinsser wrote the book *Rats, Lice and History*, alleging that both body and head lice transmit typhus between humans. Despite this, the modern view is that only the body louse can transmit the disease.

Detail showing delousing from Jan Siberechts' painting *Cour de ferme* ("Farmyard"), 1662.

Soldiers in the trenches of the First World War suffered severely from lice, and the typhus they carried. The Germans boasted that they had lice under effective control, but themselves suffered badly from lice in the Second World War on the Eastern Front, especially in the Battle of Stalingrad. "Delousing" became a grim euphemism for the extermination of Jews in concentration camps such as Auschwitz under the Nazi regime.

In the psychiatric disorder delusional parasitosis, patients express a persistent irrational fear of animals such as lice and mites, imagining that they are continually infested and complaining of itching, with "an unshakable false belief that live organisms are present in the skin".

In Literature and Folklore

MOTHER LOUSE

Mother Louse, a notorious Alewife in Oxford during the mid 18th century. Her crest includes 3 lice. Image by David Loggan.

James Joyce's 1939 book *Finnegans Wake* has the character Shem the Penman infested with "foxtrotting fleas, the lieabed lice, ... bats in his belfry".

Clifford E. Trafzer's *A Chemehuevi Song: The Resilience of a Southern Paiute Tribe* retells the story of Sinawavi (Coyote)'s love for Poowavi (Louse). Her eggs are sealed in a basket woven by her mother, who gives it to Coyote, instructing him not to open it before he reaches home. Hearing voices coming from it, however, Coyote opens the basket and the people, the world's first human beings, pour out of it in all directions.

The Irish songwriter John Lyons (b. 1934) wrote the popular song *The Kilkenny Louse House*. The song contains the lines "Well we went up the stairs and we put out the light, Sure in less than five minutes, I had to show fight. For the fleas and the bugs they collected to march, And over me stomach they formed a great arch". It has been recorded by Christie Purcell (1952), Mary Delaney on *From Puck to Appleby* (2003), and the Dubliners on *Double Dubliners* (1972) among others.

Robert Burns dedicated a poem to the Louse, inspired by witnessing one on a lady's bonnet in church: "Ye ugly, creepin, blastid wonner, Detested, shunn'd, by saint and sinner, How dare ye set your fit upon her, sae fine lady! Gae somewhere else, and seek your dinner on some poor body." John Milton in *Paradise Lost* mentioned the biblical plague of lice visited upon pharaoh: "Frogs, lice, and flies must all his palace fill with loathed intrusion, and filled all the land." John Ray recorded a Scottish proverb, "Gie a beggar a bed and he'll repay you with a Louse." In Shakespeare's *Troilus and Cressida*, Thersites compares Menelaus, brother of Agamemnon, to a louse: "Ask me not what I would be, if I were not Thersites; for I care not to be the louse of a lazar, so I were not Menelaus."

Flea

Fleas are small flightless insects that form the order **Siphonaptera**. They are external parasites of warm-blooded animals, living by hematophagy off the blood of mammals and birds. The adults are up to about 3 mm (0.12 in) long and usually brown. They are flattened sideways to enable them to move through their host's fur or feathers, and have strong claws to prevent themselves from being dislodged. They are wingless insects, with mouthparts adapted for piercing skin and sucking blood and hind legs adapted for jumping. They can leap a distance of some 50 times their body length, a feat second only to jumps made by froghoppers. The larvae are worm-like with no limbs; they have chewing mouthparts and feed on organic debris.

Over 2,500 species of fleas have been described worldwide. The Siphonaptera are most closely related to the snow scorpionflies (Boreidae), placing them within the endopterygote insect order Mecoptera.

Fleas arose in the early Cretaceous, most likely as ectoparasites of mammals and marsupials, before moving on to other groups including birds. Each species of flea is more or less a specialist on its host animal species: many species never breed on any other host, though some are less selective. Some families of fleas are exclusive to a single host group: for example, the Malacopsyllidae are found only on armadillos, the Ischnopsyllidae only on bats, and the Chimaeropsyllidae only on elephant shrews. The oriental rat flea, *Xenopsylla cheopis*, is a vector of *Yersinia pestis*, the bacterium which causes bubonic plague. The disease was spread by rodents such as the black rat, which were bitten by fleas that then infected humans. Major outbreaks included the Plague of Justinian and the Black Death, both of which killed a sizeable fraction of the world's population.

Fleas appear in human culture in such diverse forms as flea circuses, poems like John Donne's erotic *The Flea*, works of music such as by Modest Mussorgsky, and a film by Charlie Chaplin.

Morphology and Behavior

Fleas are wingless insects, 1/16 to 1/8-inch (1.5 to 3.3 mm) long, that are agile, usually dark colored (for example, the reddish-brown of the cat flea), with a proboscis, or stylet, adapted to feeding by piercing the skin and sucking their host's blood through their epipharynx. Flea legs end in strong claws that are designed to grasp a host.

Unlike other insects, fleas do not possess compound eyes but instead only have simple eyespots with a single biconvex lens; some species lack eyes altogether. Their bodies are laterally compressed, permitting easy movement through the hairs or feathers on the host's body (or in the case of humans, under clothing). The flea body is covered with hard plates called sclerites. These sclerites are covered with many hairs and short spines directed backward, which also assist its movements on the host. The tough body is able to withstand great pressure, likely an adaptation to survive attempts to eliminate them by scratching.

Fleas lay tiny, white, oval eggs. The larvae are small and pale, have bristles covering their worm-like bodies, lack eyes, and have mouth parts adapted to chewing. The larvae feed on organic matter, especially the feces of mature fleas, which contain dried blood. Adults feed only on fresh blood.

Jumping

Their legs are long, the hind pair well adapted for jumping; a flea can jump vertically up to 7 in (18 cm) and horizontally up to 13 in (33 cm),making the flea one of the best jumpers of all known animals (relative to body size), second only to the froghopper. The flea jump is so rapid and forceful that it exceeds the capabilities of muscle, and instead of relying on direct muscle power, fleas store muscle energy in a pad of the elastic protein named resilin before releasing it rapidly (like a human using a bow and arrow). Immediately before the jump, muscles contract and deform the resilin pad, slowly storing energy which can then be released extremely rapidly to power leg extension for propulsion. To prevent premature release of energy or motions of the leg, the flea employs a "catch mechanism". Early in the jump, the tendon of the primary jumping muscle passes slightly behind the coxa-trochanter joint, generating a torque which holds the joint closed with the leg close to the body. To trigger jumping, another muscle pulls the tendon forward until it passes the joint axis, generating the opposite torque to extend the leg and power the jump by release of stored energy.

Researchers with the University of Cambridge in England found that fleas take off from their tibiae and tarsi (the insect equivalent of feet) and not their trochantera, or knees.

Life cycle and Development

Fleas are holometabolous insects, going through the four lifecycle stages of egg, larva, pupa, and imago (adult). In most species, neither female nor male fleas are fully mature when they first emerge but must feed on blood before they become capable of reproduction. The first blood meal triggers the maturation of the ovaries in females and the dissolution of the testicular plug in males, and copulation soon follows. Some species breed all year round while others synchronise their activities with their hosts' life cycles or with local environmental factors and climatic conditions. Flea populations consist of roughly 50% eggs, 35% larvae, 10% pupae, and 5% adults.

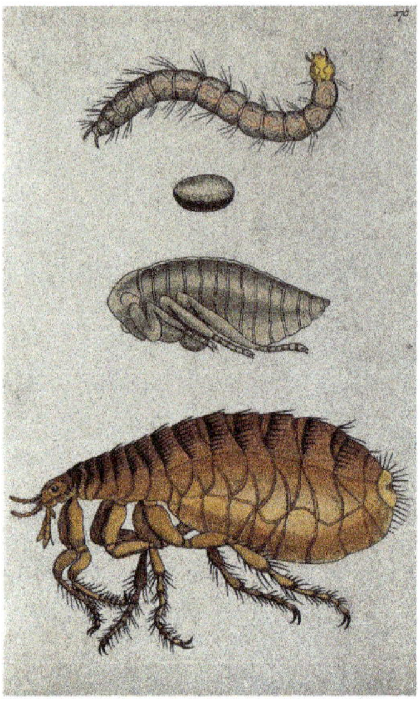

Dog flea (from top) larva, egg, pupa and adult.

Egg

The number of eggs laid depends on species, with batch sizes ranging from two to several dozen. The total number of eggs produced in a female's lifetime (fecundity) varies from around one hundred to several thousand. In some species, the flea lives in the nest or burrow and the eggs are deposited on the substrate, but in others, the eggs are laid on the host itself and can easily fall off onto the ground. Because of this, areas where the host rests and sleeps become one of the primary habitats of eggs and developing larvae. The eggs take around two days to two weeks to hatch. Experiments have shown that fleas lay more eggs on hosts which have limited food intakes, and that eggs and larvae survive better under these conditions, perhaps because the host's immune system is compromised.

Larva

Flea larvae emerge from the eggs to feed on any available organic material such as dead insects, faeces, conspecific eggs, and vegetable matter. In laboratory studies, some dietary diversity seems necessary for proper larval development. Blood-only diets allow only 12% of larvae to mature, whereas blood and yeast or dog chow diets allow almost all larvae to mature. Another study also showed that 90% of larvae matured into adults when the diet included nonviable eggs. They are blind and avoid sunlight, keeping to dark, humid places such as sand or soil, cracks and crevices, under carpets and in bedding.

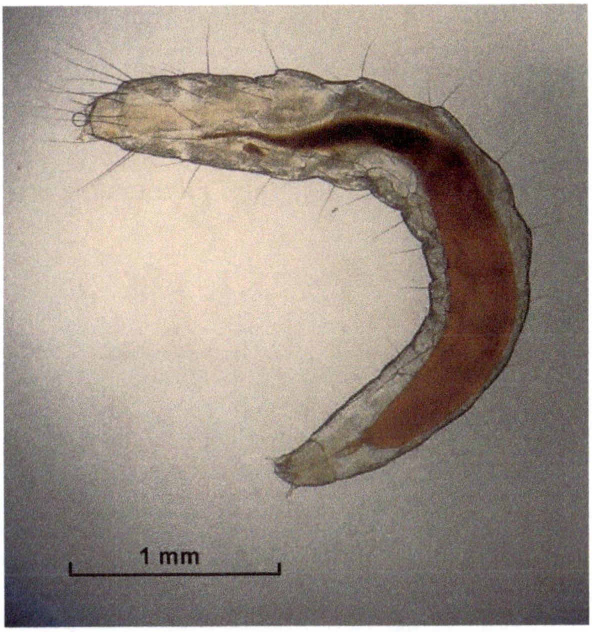

Flea larva.

Pupa

Given an adequate supply of food, larvae pupate and weave silken cocoons after three larval stages. Within the cocoon, the larva moults for a final time and undergoes metamorphosis into the adult form. This can take just four days, but may take much longer under adverse conditions, and there follows a variable-length stage during which the pre-emergent adult awaits a suitable opportunity to emerge. Trigger factors for emergence include vibrations (including sound), heat (in warm-blooded hosts), and increased levels of carbon dioxide, all of which stimuli may indicate the presence of a suitable host. Large numbers of pre-emergent fleas may be present in otherwise flea-free environments, and the introduction of a suitable host may trigger a mass emergence.

Adult

Once the flea reaches adulthood, its primary goal is to find blood and then to reproduce. Female fleas can lay 5000 or more eggs over their life, permitting rapid increase in numbers. Generally speaking, an adult flea only lives for 2 or 3 months. Without a host to provide a blood meal, a flea's life can be as short as a few days. Under ideal conditions of temperature, food supply, and humidity, adult fleas can live for up to a year and a half. Completely developed adult fleas can live for several months without eating, so long as they do not emerge from their puparia. Optimum temperatures for the flea's life cycle are 21 °C to 30 °C (70 °F to 85 °F) and optimum humidity is 70%.

Adult female rabbit fleas, *Spilopsyllus cuniculi*, can detect the changing levels of cortisol and corticosterone hormones in the rabbit's blood that indicate it is getting close to giving birth. This triggers sexual maturity in the fleas and they start producing eggs. As soon as the baby rabbits are born, the fleas make their way down to them and once on board they start feeding, mating, and laying eggs. After 12 days, the adult fleas make their way back to the mother. They complete this mini-migration every time she gives birth.

Taxonomy and Phylogeny

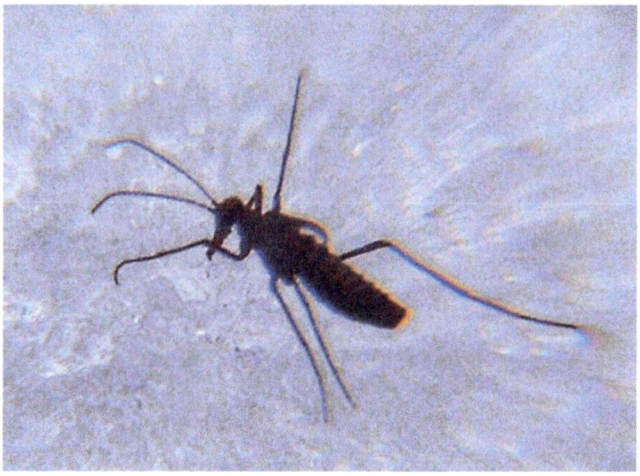

The snow scorpionflies (Boreidae) are the sister clade to the Siphonaptera.

Between 1735 and 1758, the Swedish naturalist Carl Linnaeus first classified insects, doing so on the basis of their wing structure. One of the seven orders into which he divided them was "Aptera", meaning wingless, a group in which as well as fleas, he included spiders, woodlice and myriapods. It wasn't until 1810 that the French zoologist Pierre André Latreille reclassified the insects on the basis of their mouthparts as well as their wings, splitting Aptera into Thysanura (silverfish), Anoplura (sucking lice) and Siphonaptera (fleas), at the same time separating off the arachnids and crustaceans into their own subphyla. The group's name, Siphonaptera, is zoological Latin from the Greek *siphon* (a tube) and *aptera* (wingless).

Fleas are related to the Diptera (true flies) and the Mecoptera (scorpion flies) as shown in the cladogram, based on a 2008 analysis of four loci (18S and 28S ribosomal DNA, cytochrome oxidase II, and elongation factor 1-alpha) for 128 flea taxa from around the world. The Boreidae (snow scorpionflies) are the sister clade to the Siphonaptera.

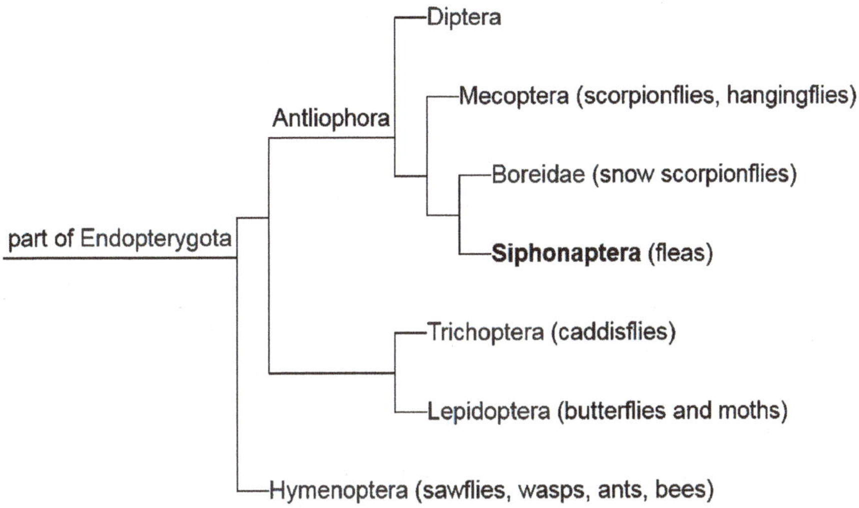

Cenozoic flea in amber, c. 20 mya, is morphologically modern.

Fossils of wingless "pre-fleas" with siphonate (sucking) mouthparts from the middle Jurassic to early Cretaceous have been found in northeastern China. These belonged to three proposed extinct families, the Pseudopulicidae, the Saurophthiridae, and the Tarwiniidae.

The last common ancestor of modern Siphonaptera separated from the Mecoptera during the early Cretaceous. Most flea families formed after the end of the Cretaceous (in the Paleogene and onwards). Fleas probably arose in the southern continental area of Gondwana, and migrated rapidly northwards from there. They most likely evolved with mammal and marsupial hosts, only later moving to birds and monotremes.

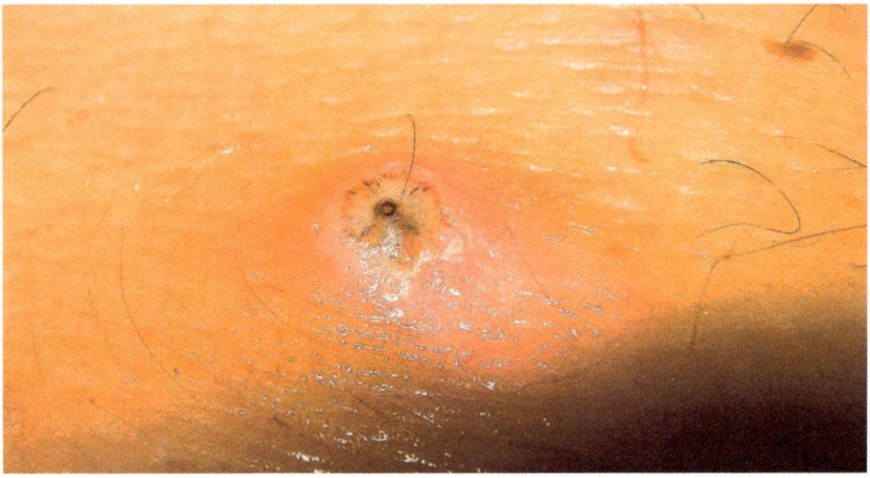

Jigger, *Tunga penetrans*, in human skin.

Siphonaptera is a relatively small order of insects: members of the order undergo complete metamorphosis and are secondarily wingless (their ancestors had wings which modern forms have lost). In 2005, Medvedev listed 2005 species in 242 genera, and despite subsequent descriptions of new species, bringing the total up to around 2500 species, this is the most complete database available. The order is divided into four infraorders and eighteen families. Some families are exclusive to a single host group; these include the Malacopsyllidae (armadillos), Ischnopsyllidae (bats) and Chimaeropsyllidae (elephant shrews).

Many of the known species are little studied. Some 600 species (a quarter of the total) are known from a single record from a single host. Over 94% of species are associated with mammalian hosts, and only about 3% of species can be considered to be specific parasites of birds. The fleas on birds are thought to have originated from mammalian fleas; at least sixteen separate groups of fleas switched to avian hosts during the evolutionary history of the Siphonaptera. Occurrences of fleas on reptiles is accidental, and fleas have been known to feed on the hemolymph (bloodlike body fluid) of ticks.

Flea phylogeny was long neglected, the discovery of homologies with the parts of other insects being made difficult by their extreme specialization. Whiting and colleagues prepared a detailed molecular phylogeny in 2008, with the basic structure shown in the cladogram. The Tungidae, including the harmful chigoe flea or jigger, is sister to the rest of the Siphonaptera.

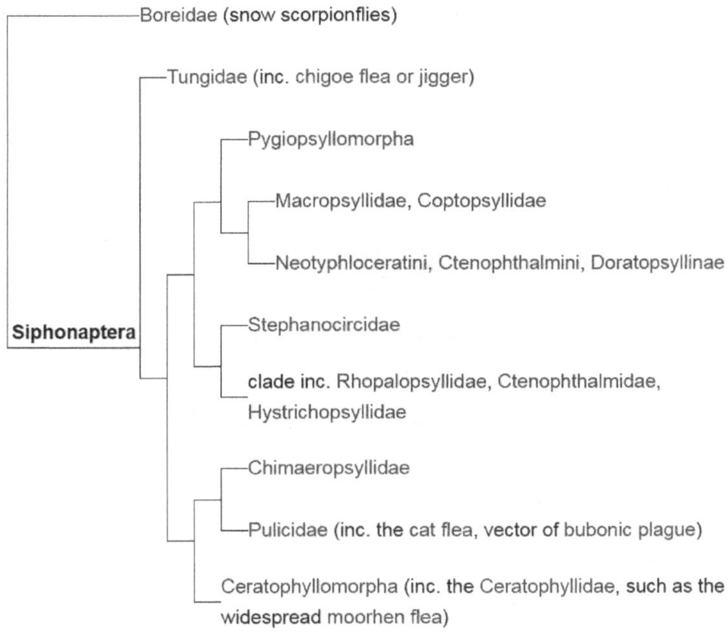

Relationship with Host

Fleas feed on a wide variety of warm-blooded vertebrates including humans, dogs, cats, rabbits, squirrels, ferrets, rats, mice and birds. Fleas normally specialise in one host species or group of species, but can often feed but not reproduce on other species. *Ceratophyllus gallinae* affects poultry as well as wild birds. As well as the degree of relatedness of a potential host to the flea's original host, it has been shown that avian fleas that exploit a range of hosts, only parasitise species with low immune responses. In general, host specificity decreases as the size of the host species decreases. Another factor is the opportunities available to the flea to change host species; this is smaller in colonially nesting birds, where the flea may never encounter another species, than it is in solitary nesting birds. A large, long-lived host provides a stable environment that favours host-specific parasites.

One theory of human hairlessness is that the loss of hair helped humans to reduce their burden of fleas and other ectoparasites.

Direct Effects of Bites

In many species, fleas are principally a nuisance to their hosts, causing an itching sensation which in turn causes the host to try to remove the pest by biting, pecking or scratching. Fleas are not simply a source of annoyance, however. Flea bites cause a slightly raised, swollen itching spot to form; this has a single puncture point at the centre, like a mosquito bite. Besides this, the eczematous itchy skin disease flea allergy dermatitis is common in many host species, including dogs and cats. The bites often appear in clusters or lines of two bites, and can remain itchy and inflamed for up to several weeks afterwards. Fleas can lead to hair loss as a result of frequent scratching and biting by the animal, and can cause anemia in extreme cases.

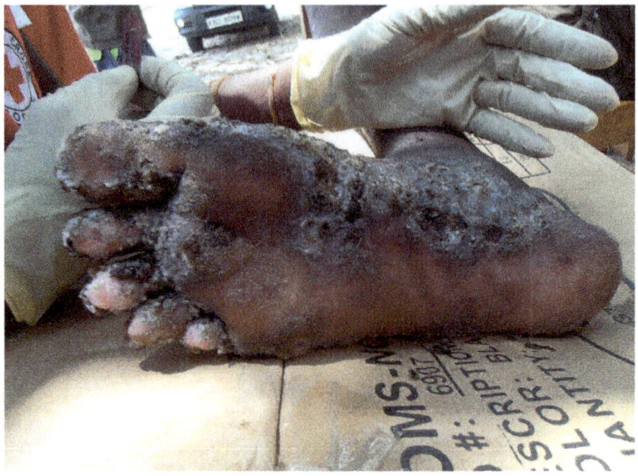

Human foot infested with jigger fleas, *Tunga penetrans*.

As a Vector

Fleas are vectors for viral, bacterial and rickettsial diseases of humans and other animals, as well as of protozoan and helminth parasites.Bacterial diseases carried by fleas include murine or endemic typhus. and bubonic plague. Fleas can transmit *Rickettsia typhi*, *Rickettsia felis*, and *Bartonella henselae*, and the myxomatosis virus. They can carry *Hymenolepiasis* tapeworms and Trypanosome protozoans. The chigoe flea or jigger (*Tunga penetrans*) causes the disease tungiasis, a major public health problem around the world. Fleas that specialize as parasites on specific mammals may use other mammals as hosts; thus, humans may be bitten by cat and dog fleas.

Relationship with Humans

In literature and Art

Fleas have appeared in poetry, literature, music and art; these include Robert Hooke's drawing of a flea under the microscope in his pioneering book *Micrographia* published in 1665, poems by Donne and Jonathan Swift, works of music by Giorgio Federico Ghedini and Modest Mussorgsky, a play by Georges Feydeau, a film by Charlie Chaplin, and paintings by artists such as Giuseppe Crespi, Giovanni Battista Piazzetta, and Georges de La Tour.

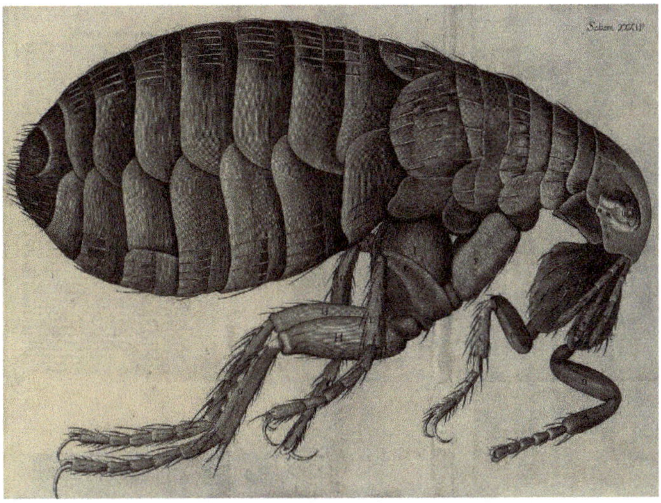

Robert Hooke's drawing of a flea in *Micrographia*, 1665.

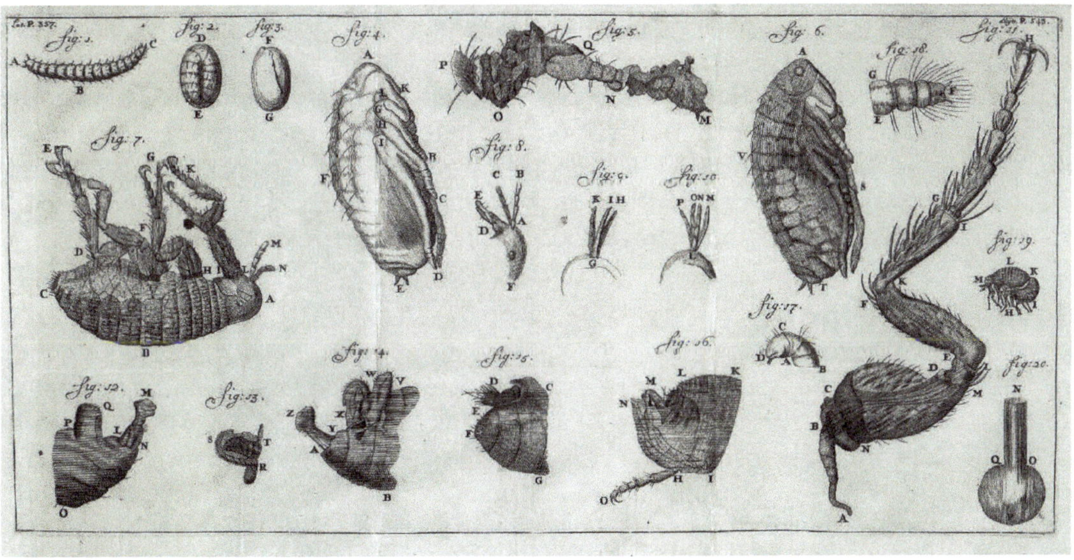

Development of the flea from egg to adult. Antonie van Leeuwenhoek, c. 1680.

John Donne's erotic metaphysical poem "The Flea", published in 1633 after his death, uses the conceit of a flea, which has sucked blood from the male speaker and his female lover, as an extended metaphor for their sexual relationship. The speaker tries to convince a lady to sleep with him, arguing that if the mingling of their blood in the flea is innocent, then sex would be also.

Flea Circuses

A flea circus: "The Go-As-You-Please Race, as seen through a Magnifying Glass", engraved by J. G. Francis, from an article by C. F. Holder in St. Nicholas Magazine, 1886.

Flea circuses provided entertainment to nineteenth century audiences. These circuses, extremely popular in Europe from 1830 onwards, featured fleas dressed as humans or towing miniature carts, chariots, rollers or cannon. These devices were originally made by watchmakers or jewellers to show off their skill at miniaturization. A ringmaster called a "professor" accompanied their performance with a rapid circus patter.

Carriers of Plague

Oriental rat fleas, *Xenopsylla cheopis*, can carry the coccobacillus *Yersinia pestis.* The infected fleas feed on rodent vectors of this bacterium, such as the black rat, *Rattus rattus*, and then infect human populations with the plague, as has happened repeatedly from ancient times, as in the Plague of Justinian in 541–542. Outbreaks killed up to 200 million people across Europe between 1346 and 1671. The Black Death pandemic between 1346 and 1353 likely killed over a third of the population of Europe.

Because fleas carry plague, they have seen service as a biological weapon. During World War II, the Japanese army dropped fleas infested with *Y. pestis* in China. The bubonic and septicaemic plagues are the most probable form of the plague that would spread as a result of a bioterrorism attack that used fleas as a vector.

The Great Plague of London, in 1665, killed up to 100,000 people.

The Rothschild Collection

The banker Charles Rothschild devoted much of his time to entomology, creating a large collection of fleas now in the Rothschild Collection at the Natural History Museum, London. He discovered and named the plague vector flea, *Xenopsylla cheopis*, also known as the oriental rat flea, in 1903. Using what was probably the world's most complete collection of fleas of about 260,000 specimens (representing some 73% of the 2,587 species and subspecies so far described), he described around 500 species and subspecies of Siphonaptera. He was followed in this interest by his daughter Miriam Rothschild, who helped to catalogue his enormous collection of the insects in seven volumes.

Flea Treatments

Fleas have a significant economic impact. In America alone, approximately $2.8 billion is spent annually on flea-related veterinary bills and another $1.6 billion annually for flea treatment with pet groomers. Four billion dollars is spent annually for prescription flea treatment and $348 million for flea pest control.

BED BUG

Bed bugs are parasitic insects of the cimicid family that feed exclusively on blood. ***Cimex lectularius***, the common bed bug, is the best known as it prefers to feed on human blood. Other *Cimex* species specialize in other animals, e.g., bat bugs, such as *Cimex pipistrelli* (Europe), *Cimex pilosellus* (western US), and *Cimex adjunctus* (entire eastern US).

The name bed bug derives from the preferred habitat of *Cimex lectularius*: warm houses and especially near or inside beds and bedding or other sleep areas. Bed bugs are mainly active at night, but are not exclusively nocturnal. They usually feed on their hosts without being noticed.

A number of adverse health effects may result from bed bug bites, including skin rashes, psychological effects, and allergic symptoms. Bed bugs are not known to transmit any pathogens as disease vectors. Certain signs and symptoms suggest the presence of bed bugs; finding the adult insects confirms the diagnosis.

Bed bugs have been known as human parasites for thousands of years. At a point in the early 1940s, they were mostly eradicated in the developed world, but have increased in prevalence since 1995, likely due to pesticide resistance, governmental bans on effective pesticides, and international travel. Because infestation of human habitats has begun to increase, bed bug bites and related conditions have been on the rise as well.

Infestation

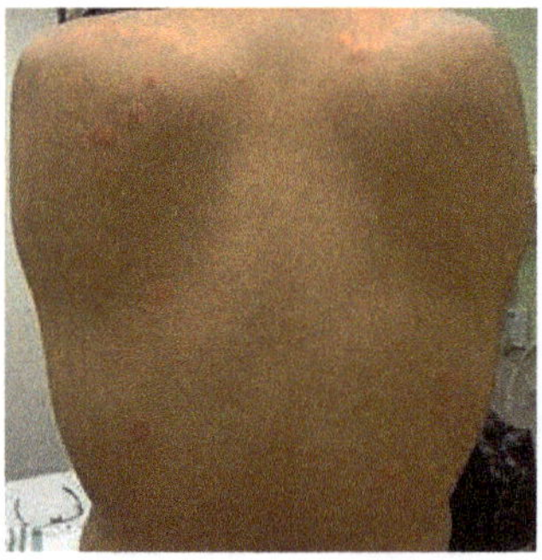

Bedbug bites.

Diagnosis of an infestation involves both finding bed bugs and the occurrence of compatible symptoms. Treatment involves the elimination of the insect (including its eggs) and taking measures to treat symptoms until they resolve.

Bed bug bites or cimicosis may lead to a range of skin manifestations from no visible effects to prominent blisters. Effects include skin rashes, psychological effects, and allergic symptoms.

Although bed bugs can be infected with at least 28 human pathogens, no studies have found that the insects are capable of transmitting any of these to humans. They have been found with methicillin-resistant *Staphylococcus aureus* (MRSA)and with vancomycin-resistant *Enterococcus faecium* (VRE), but the significance of this is still unknown.

Investigations into potential transmission of HIV, MRSA, hepatitis B, hepatitis C, and hepatitis E have not shown that bed bugs can spread these diseases. However, arboviruses may be transmissible.

Description

Physical

Adult bed bugs are light brown to reddish-brown, flattened, oval-shaped, and have no hind wings. The front wings are vestigial and reduced to pad-like structures. Bed bugs have segmented abdomens with microscopic hairs that give them a banded appearance. Adults grow to 4–5 mm (0.16–0.20 in) long and 1.5–3 mm (0.059–0.118 in) wide.

Newly hatched nymphs are translucent, lighter in color, and become browner as they moult and reach maturity. A bed bug nymph of any age that has just consumed a blood meal has a bright red, translucent abdomen, fading to brown over the next several hours, and to opaque black within two days as the insect digests its meal. Bed bugs may be mistaken for other insects, such as booklice, small cockroaches, or carpet beetles; however, when warm and active, their movements are more ant like and, like most other true bugs, they emit a characteristic disagreeable odor when crushed.

Bed bugs use pheromones and kairomones to communicate regarding nesting locations, feeding, and reproduction.

The lifespan of bed bugs varies by species and is also dependent on feeding.

Bed bugs can survive a wide range of temperatures and atmospheric compositions. Below 16.1 °C (61.0 °F), adults enter semihibernation and can survive longer; they can survive for at least five days at −10 °C (14 °F), but die after 15 minutes of exposure to −32 °C (−26 °F). Common commercial and residential freezers reach temperatures low enough to kill most life stages of bed bug, with 95% mortality after 3 days at −12 °C (10 °F). They show high desiccation tolerance, surviving low humidity and a 35–40 °C range even with loss of one-third of body weight; earlier life stages are more susceptible to drying out than later ones.

The thermal death point for *C. lectularius* is 45 °C (113 °F); all stages of life are killed by 7 minutes of exposure to 46 °C (115 °F). Bed bugs apparently cannot survive high concentrations of carbon dioxide for very long; exposure to nearly pure nitrogen atmospheres, however, appears to have relatively little effect even after 72 hours.

Feeding Habits

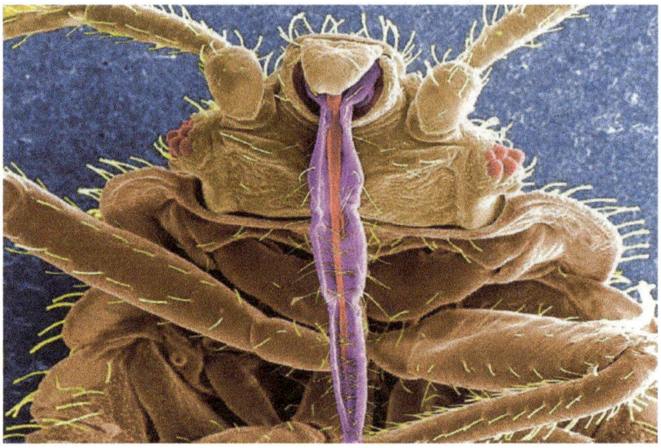

Scanning electron micrograph (SEM) digitally colorized with skin-piercing mouthparts highlighted in purple and red

Bed bugs are obligatory hematophagous (bloodsucking) insects. Most species feed on humans only when other prey are unavailable. They obtain all the additional moisture they need from water vapor in the surrounding air. Bed bugs are attracted to their hosts primarily by carbon dioxide, secondarily by warmth, and also by certain chemicals. Bedbugs prefer exposed skin, preferably the face, neck, and arms of a sleeping person.

Bedbugs have mouth parts that saw through the skin, and inject saliva with anticoagulants and painkillers. Sensitivity of humans varies from extreme allergic reaction to no reaction at all (about 20%). The bite usually produces a swelling with no red spot, but when many bugs feed on a small area, reddish spots may appear after the swelling subsides.

Although under certain cool conditions adult bed bugs can live for over a year without feeding, under typically warm conditions they try to feed at five- to ten-day intervals, and adults can survive for about five months without food. Younger instars cannot survive nearly as long, though even the vulnerable newly hatched first instars can survive for weeks without taking a blood meal.

At the 57th annual meeting of the Entomological Society of America in 2009, newer generations of pesticide-resistant bed bugs in Virginia were reported to survive only two months without feeding.

DNA from human blood meals can be recovered from bed bugs for up to 90 days, which mean they can be used for forensic purposes in identifying on whom the bed bugs have fed.

Feeding Physiology

A bed bug pierces the skin of its host with a stylet fascicle, rostrum, or "beak". The rostrum is composed of the maxillae and mandibles, which have been modified into elongated shapes from a basic, ancestral style. The right and left maxillary stylets are connected at their midline and a section at the centerline forms a large food canal and a smaller salivary canal. The entire maxillary and mandibular bundle penetrates the skin.

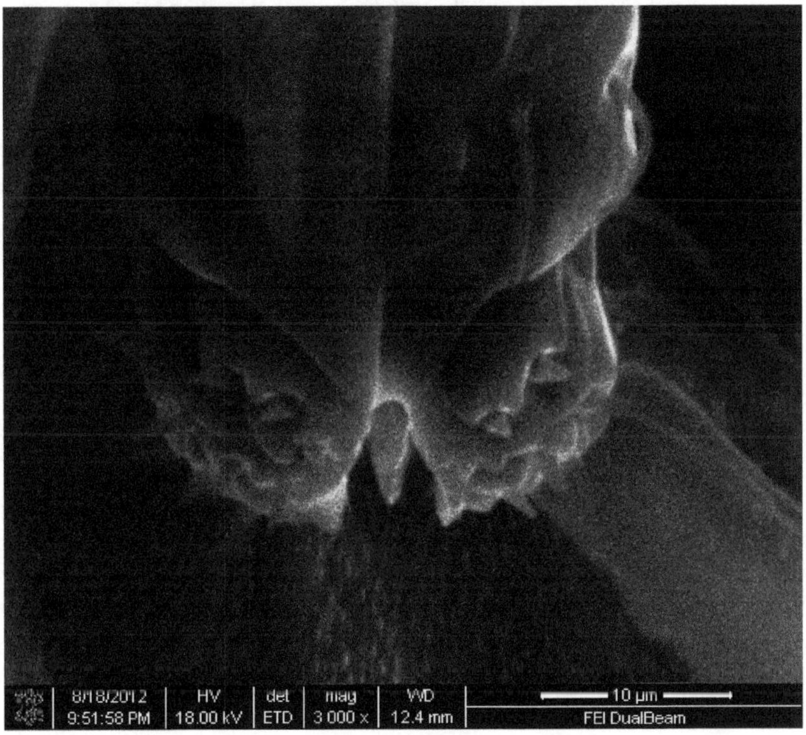

Rostrum tip.

The tips of the right and left maxillary stylets are not the same; the right is hook-like and curved, and the left is straight. The right and left mandibular stylets extend along the outer sides of their respective maxillary stylets and do not reach anywhere near the tip of the fused maxillary stylets. The stylets are retained in a groove in the labium, and during feeding, they are freed from the groove as the jointed labium is bent or folded out of the way; its tip never enters the wound.

The mandibular stylet tips have small teeth, and through alternately moving these stylets back and forth, the insect cuts a path through tissue for the maxillary bundle to reach an appropriately sized blood vessel. Pressure from the blood vessel itself fills the insect with blood in three to five minutes. The bug then withdraws the stylet bundle from the feeding position and retracts it back into the labial groove, folds the entire unit back under the head, and returns to its hiding place. It takes between five and ten minutes for a bed bug to become completely engorged with blood. In all, the insect may spend less than 20 minutes in physical contact with its host, and does not try to feed again until it has either completed a moult or, if an adult, has thoroughly digested the meal.

Reproduction

Male bed bug traumatically inseminating a female.

All bed bugs mate by traumatic insemination. Female bed bugs possess a reproductive tract that functions during oviposition, but the male does not use this tract for sperm insemination. Instead, the male pierces the female's abdomen with his hypodermic penis and ejaculates into the body cavity. In all bed bug species except *Primicimex cavernis,* sperm are injected into the mesospermalege, a component of the spermalege, a secondary genital structure that reduces the wounding and immunological costs of traumatic insemination. Injected sperm travel via the haemolymph (blood) to sperm storage structures called seminal conceptacles, with fertilisation eventually taking place at the ovaries.

Male bed bugs sometimes attempt to mate with other males and pierce their abdomens. This behaviour occurs because sexual attraction in bed bugs is based primarily on size, and males mount any freshly fed partner regardless of sex. The "bed bug alarm pheromone" consists of (*E*)-2-octenal and (*E*)-2-hexenal. It is released when a bed bug is disturbed, as during an attack by a predator. A 2009 study demonstrated the alarm pheromone is also released by male bed bugs to repel other males that attempt to mate with them.

Cimex lectularius and *C. hemipterus* mate with each other given the opportunity, but the eggs then produced are usually sterile. In a 1988 study, one of 479 eggs was fertile and resulted in a hybrid, *Cimex hemipterus* × *lectularius*.

Sperm Protection

Cimex lectularius males have environmental microbes on their genitals. These microbes damage sperm cells, leaving them unable to fertilize female gametes. Due to these dangerous microbes, males have evolved antimicrobial ejaculate substances that prevent sperm damage. When the microbes contact sperm or the male genitals, the bed bug releases antimicrobial substances. Many species of these microbes live in the bodies of females after mating. The microbes can cause infections in the females. It has been suggested that females receive benefit from the ejaculate. Though the benefit is not direct, females are able to produce more eggs than optimum increasing the amount of the females' genes in the gene pool.

Sperm and Seminal Fluid Allocation

In organisms, sexual selection extends past differential reproduction to affect sperm composition, sperm competition, and ejaculate size. Males of *C. lectularius* allocate 12% of their sperm and 19% of their seminal fluid per mating. Due to these findings, Reinhard et. al proposed that multiple mating is limited by seminal fluid and not sperm. After measuring ejaculate volume, mating rate and estimating sperm density, Reinhardt et al. showed that mating could be limited by seminal fluid. Despite these advances, the cost difference between ejaculate-dose dependence and mating frequency dependence have not been explored.

Egg Production

Males fertilize females only by traumatic insemination into the structure called the ectospermalege (the organ of Berlese, however the organ of Ribaga, as it was first named, was first designated as an organ of stridulation. These two names are not descriptive, so other terminologies are used). On fertilization, the female's ovaries finish developing, which suggests that sperm plays a role other than fertilizing the egg. Fertilization also allows for egg production through the corpus allatum. Sperm remains viable in a female's spermathecae (a better term is conceptacle), a sperm-carrying sack, for a long period of time as long as body temperature is optimum. The female lays fertilized eggs until she depletes the sperm found in her conceptacle. After the depletion of sperm, she lays a few sterile eggs. The number of eggs a *C. lectularius* female produces does not depend on the sperm she harbors, but on the female's nutritional level.

Alarm Pheromones

In *C. lectularius*, males sometimes mount other males because male sexual interest is directed at any recently fed individual regardless of their sex, but unfed females may also be mounted. Traumatic insemination is the only way for copulation to occur in bed bugs. Females have evolved the spermalege to protect themselves from wounding and infection. Because males lack this organ, traumatic insemination could leave them badly injured. For this reason, males have evolved alarm pheromones to signal their sex to other males. If a male *C. lectularius* mounts another male, the mounted male releases the pheromone signal and the male on top stops before insemination.

Females are capable of producing alarm pheromones to avoid multiple mating, but they generally do not do so. Two reasons are proposed as to why females do not release alarm pheromones to protect themselves. First, alarm pheromone production is costly. Due to egg production, females may refrain from spending additional energy on alarm pheromones. The second proposed reason is that releasing the alarm pheromone reduces the benefits associated with multiple mating. Benefits of multiple mating include material benefits, better quality nourishment or more nourishment, genetic benefits including increased fitness of offspring, and finally, the cost of resistance may be higher than the benefit of consent—which appears the case in *C. lectularius*.

Life Stages

Bed bugs have five immature nymph life stages and a final sexually mature adult stage. They shed their skins through ecdysis at each stage, discarding their outer exoskeleton, which is somewhat clear, empty exoskeletons of the bugs themselves. Bed bugs must molt six times before becoming fertile adults, and must consume at least one blood meal to complete each molt.

Each of the immature stages lasts about a week, depending on temperature and the availability of food, and the complete lifecycle can be completed in as little as two months (rather long compared to other ectoparasites). Fertilized females with enough food lay three to four eggs each day continually until the end of their lifespans (about nine months under warm conditions), possibly generating as many as 500 eggs in this time. Genetic analysis has shown that a single pregnant bed bug, possibly a single survivor of eradication, can be responsible for an entire infestation over a matter of weeks, rapidly producing generations of offspring.

Slide of *C. lectularius*

On the right is recently sloughed skin from its nymph stage

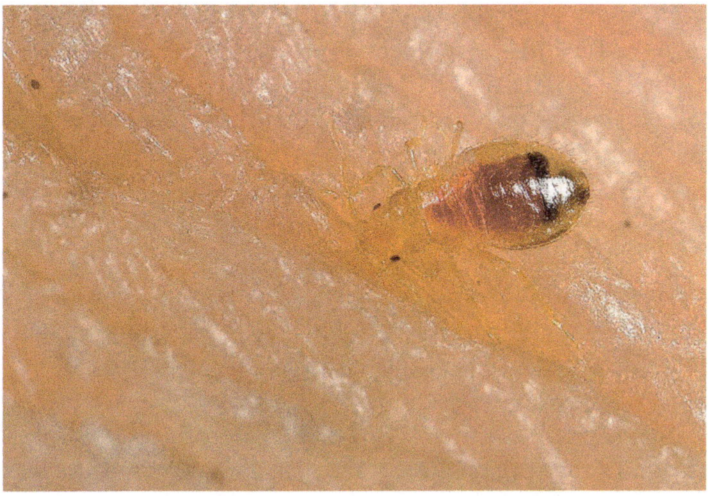

Nymph feeding on host

Blood-fed stage (note differences in color with respect to digestion of blood meal)

Sexual Dimorphism

Sexual dimorphism occurs in *C. lectularius*, with the females larger in size than the males on average. The abdomens of the sexes differ in that the males appear to have “pointed” abdomens, which are actually their copulatory organs, while females have more rounded abdomens. Since males are attracted to large body size, any bed bug with a recent blood meal can be seen as a potential mate. However, males will mount unfed, flat females on occasion. The female is able to curl her abdomen forward and underneath toward the head to not mate. Males are generally unable to discriminate between the sexes until after mounting, but before inseminating.

Host searching

C. lectularius only feeds every five to seven days, which suggests that it does not spend the majority of its life searching for a host. When a bed bug is starved, it leaves its shelter and searches for a host. If it successfully feeds, it returns to its shelter. If it does not feed, it continues to search for a host. After searching—regardless of whether or not it has eaten—the bed bug returns to the shelter to aggregate before the photophase (period of light during a day-night cycle). Reis argues that two reasons explain why *C. lectularius* would return to its shelter and aggregate after feeding. One is to find a mate and the other is to find shelter to avoid getting smashed after eating.

Aggregation and Dispersal Behavior

C. lectularius aggregates under all life stages and mating conditions. Bed bugs may choose to aggregate because of predation, resistance to desiccation, and more opportunities to find a mate. Airborne pheromones are responsible for aggregations. Another source of aggregation could be the recognition of other *C. lectularius* bugs through mechanoreceptors located on their antennae. Aggregations are formed and disbanded based on the associated cost and benefits. Females are more often found separate from the aggregation than males. Females are more likely to expand the population range and find new sites. Active female dispersal can account for treatment failures. Males, when found in areas with few females, abandon an aggregation to find a new mate. The males excrete an aggregation pheromone into the air that attracts virgin females and arrests other males.

Detection

Eggs and two adults found inside a dresser.

Detection dog.

Fecal spot.

Roaming on carpet wrinkles.

Bed bugs can exist singly, but tend to congregate once established. Though strictly parasitic, they spend only a tiny fraction of their lifecycles physically attached to hosts. Once a bed bug finishes feeding, it relocates to a place close to a known host, commonly in or near beds or couches in clusters of adults, juveniles, and eggs—which entomologists call harborage areas or simply harborages to which the insect returns after future feedings by following chemical trails. These places can vary greatly in format, including luggage, inside of vehicles, within furniture, amongst bedside clutter—even inside electrical sockets and nearby laptop computers. Bed bugs may also nest near animals that have nested within a dwelling, such as bats, birds, or rodents. They are also capable of surviving on domestic cats and dogs, though humans are the preferred host of *C. lectularius*.

Bed bugs can also be detected by their characteristic smell of rotting raspberries. Bed bug detection dogs are trained to pinpoint infestations, with a possible accuracy rate between 11% and 83%. Homemade detectors have been developed.

Management

Eradication of bed bugs frequently requires a combination of nonpesticide approaches and the occasional use of pesticides.

Mechanical approaches, such as vacuuming up the insects and heat-treating or wrapping mattresses, are effective. A combination of heat and drying treatments is most effective. An hour at a temperature of 45 °C (113 °F) or over, or two hours at less than −17 °C (1 °F) kills them; a domestic clothes drier or steam kills bedbugs. Another study found 100% mortality rates for bed bugs exposed to temperatures greater than 50 °C (122 °F) for more than 2 minutes. Starving them is difficult as they can survive without eating for 100 to 300 days, depending on temperature. For public health reasons, individuals are encouraged to call a professional pest control service to eradicate bed bugs in a home, rather than attempting to do it themselves, particularly if they live in a multifamily building.

As of 2012, no truly effective pesticides were available. Pesticides that have historically been found effective include pyrethroids, dichlorvos, and malathion. Resistance to pesticides has increased significantly over time, and harm to health from their use is of concern. The carbamate insecticide propoxur is highly toxic to bed bugs, but it has potential toxicity to children exposed to it, and the US Environmental Protection Agency has been reluctant to approve it for indoor use. Boric acid, occasionally applied as a safe indoor insecticide, is not effective against bed bugs because they do not groom. The fungus *Beauveria bassiana* is being researched as of 2012 for its ability to control bed bugs. As bed bugs continue to adapt pesticide resistance, researchers have examined on the insect's genome to see how the adaptations develop and to look for potential vulnerabilities that can be exploited in the growth and development phases.

Predators

Natural enemies of bed bugs include the masked hunter insect (also known as "masked bed bug hunter"),cockroaches, ants, spiders (particularly *Thanatus flavidus*), mites, and centipedes (particularly the house centipede *Scutigera coleoptrata*). However, biological pest control is not considered practical for eliminating bed bugs from human dwellings.

Epidemiology

Bed bugs occur around the world. Rates of infestations in developed countries, while decreasing from the 1930s to the 1980s, have increased dramatically since the 1980s. Previously, they were common in the developing world, but rare in the developed world. The increase in the developed world may have been caused by increased international travel, resistance to insecticides, and the use of new pest-control methods that do not affect bed bugs.

The exact causes of this resurgence remain unclear; it is variously ascribed to greater foreign travel, increased immigration from the developing world to the developed world, more frequent exchange of second-hand furnishings among homes, a greater focus on control of other pests, resulting in neglect of bed bug countermeasures, and increasing resistance to pesticides. Declines in household cockroach populations that have resulted from the use of insecticides effective against this major bed bug predator have aided the bed bugs' resurgence, as have bans on DDT and other potent pesticides.

The fall in bed bug populations after the 1930s in the developed world is believed partly due to the use of DDT to kill cockroaches. The invention of the vacuum cleaner and simplification of furniture design may have also played a role. Others believe it might simply be the cyclical nature of the organism.

The common bed bug (*C. lectularius*) is the species best adapted to human environments. It is found in temperate climates throughout the world. Other species include *Cimex hemipterus*, found in tropical regions, which also infests poultry and bats, and *Leptocimex boueti*, found in the tropics of West Africa and South America, which infests bats and humans. *Cimex pilosellus* and *Cimex pipistrella* primarily infest bats, while *Haematosiphon inodora*, a species of North America, primarily infests poultry.

In November 2016, a media report noted that tropical bed bugs, *Cimex hemipterus*, which had been extirpated from the state during World War II, were discovered in Brevard County, Florida and are expected to spread in distribution within the United States.

History

C. lectularius may have originated in the Middle East in caves inhabited by bats and humans.

Bed bugs were mentioned in ancient Greece as early as 400 BC, and were later mentioned by Aristotle. Pliny's *Natural History*, first published *circa* AD 77 in Rome, claimed bed bugs had medicinal value in treating ailments such as snake bites and ear infections. (Belief in the medicinal use of bed bugs persisted until at least the 18th century, when Guettard recommended their use in the treatment of hysteria.)

Bed bugs were first mentioned in Germany in the 11th century, in France in the 13th century, and in England in 1583,though they remained rare in England until 1670. Some in the 18th century believed bed bugs had been brought to London with supplies of wood to rebuild the city after the Great Fire of London (1666). Giovanni Antonio Scopoli noted their presence in Carniola (roughly equivalent to present-day Slovenia) in the 18th century.

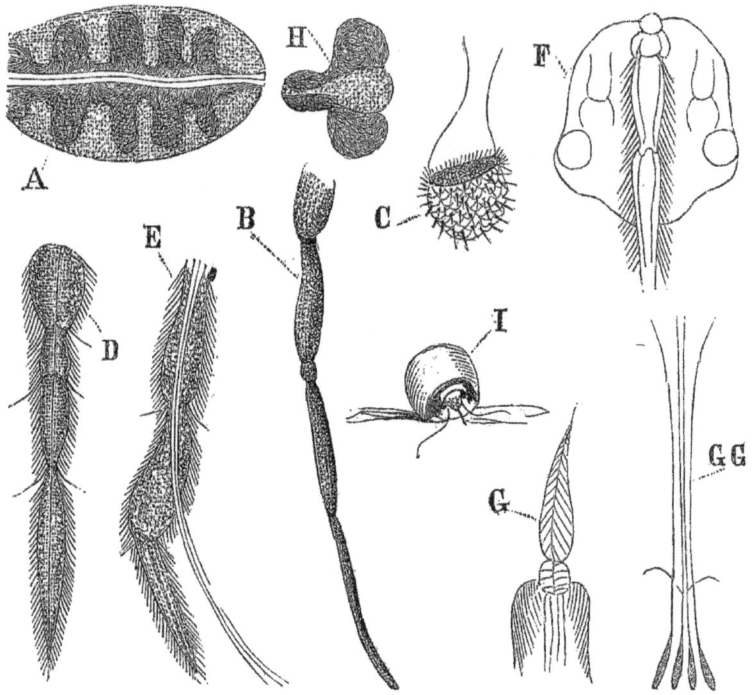

1860 engraving of bed bug parts: *A*. Intestines – *B*. Antenna of the male – *C*. Eye – *D*. Haustellum, or sucker, closed – *E*. Side view of sucker – *F*. Under part of head – *G*. Under lip – *GG*. Hair of the tube, and outside cases – *H*. Egg-bag – *I*. Larva emerging from the eggs.

Traditional methods of repelling and/or killing bed bugs include the use of plants, fungi, and insects (or their extracts), such as black pepper; black cohosh (*Actaea racemosa*); *Pseudarthria hookeri*; *Laggera alata* (Chinese *yángmáo cǎo* | 羊毛草); *Eucalyptus saligna* oil; henna (*Lawsonia inermis* or camphire); "infused oil of *Melolontha vulgaris*» (presumably cockchafer); fly agaric (*Amanita muscaria*); *Actaea* spp. (e.g. black cohosh); tobacco; "heated oil of Terebinthina" (i.e. true turpentine); wild mint (*Mentha arvensis*); narrow-leaved pepperwort (*Lepidium ruderale*); *Myrica* spp. (e.g. bayberry); Robert geranium (*Geranium robertianum*); bugbane (*Cimicifuga* spp.); "herb and seeds of *Cannabis*"; "opulus" berries (possibly maple or European cranberrybush); masked hunter bugs (*Reduvius personatus*), "and many others".

In the mid-19th century, smoke from peat fires was recommended as an indoor domestic fumigant against bed bugs.

Dusts have been used to ward off insects from grain storage for centuries, including plant ash, lime, dolomite, certain types of soil, and diatomaceous earth or Kieselguhr. Of these, diatomaceous earth in particular has seen a revival as a nontoxic (when in amorphous form) residual pesticide for bed bug abatement. While diatomaceous earth performed poorly, silica gel may be effective.

Basket-work panels were put around beds and shaken out in the morning in the UK and in France in the 19th century. Scattering leaves of plants with microscopic hooked hairs around a bed at night, then sweeping them up in the morning and burning them, was a technique reportedly used in Southern Rhodesia and in the Balkans.

Bean leaves have been used historically to trap bedbugs in houses in Eastern Europe. The trichomes on the bean leaves capture the insects by impaling the feet (tarsi) of the insects. The leaves are then destroyed.

20th century

Prior to the mid-20th century, bed bugs were very common. According to a report by the UK Ministry of Health, in 1933, all the houses in many areas had some degree of bed bug infestation. The increase in bed bug populations in the early 20th century has been attributed to the advent of electric heating, which allowed bed bugs to thrive year-round instead of only in warm weather.

Bed bugs were a serious problem at US military bases during World War II. Initially, the problem was solved by fumigation, using Zyklon Discoids that released hydrogen cyanide gas, a rather dangerous procedure. Later, DDT was used to good effect as a safer alternative.

The decline of bed bug populations in the 20th century is often credited to potent pesticides that had not previously been widely available. Other contributing factors that are less frequently mentioned in news reports are increased public awareness and slum clearance programs that combined pesticide use with steam disinfection, relocation of slum dwellers to new housing, and in some cases also follow-up inspections for several months after relocated tenants moved into their new housing.

Resurgence

Bed bug infestations resurged since the 1980s for reasons that are not clear, but contributing factors may be complacency, increased resistance, bans on pesticides, and increased international travel. The U.S. National Pest Management Association reported a 71% increase in bed bug calls between 2000 and 2005. The number of reported incidents in New York City alone rose from 500 in 2004 to 10,000 in 2009. In 2013, Chicago was listed as the number 1 city in the United States with the worst bed bug infestation. As a result, the Chicago City Council passed a bed bug control ordinance to limit their spread. Additionally, bed bugs are reaching places in which they never established before, such as southern South America.

One recent theory about bed bug reappearance in the US is that they never truly disappeared, but may have been forced to alternative hosts. Consistent with this is the finding that bed bug DNA shows no evidence of an evolutionary bottleneck. Furthermore, investigators have found high populations of bed bugs at poultry facilities in Arkansas. Poultry workers at these facilities may be spreading bed bugs, unknowingly carrying them to their places of residence and elsewhere after leaving work.

Society and Culture

"Good night, sleep tight, don't let the bed bugs bite", is a saying some people recite before they go to sleep.

Bed bug secretions can inhibit the growth of some bacteria and fungi; antibacterial components from the bed bug could be used against human pathogens, and be a source of pharmacologically active molecules as a resource for the discovery of new drugs.

Etymology

The word *bug* and its earlier spelling *bugge* originally meant "bed bug". Many other creatures are now called "bugs", such as the "ladybug" ("ladybird" outside North America) and the "potato bug"; the word is used informally for any insect, or even microscopic germs or diseases caused by these germs, but the earliest recorded use of the actual word "bug" referred to a bed bug.

The term "bed bug" may also be spelled "bedbug" or "bed-bug", though published sources consistently use the unhyphenated two-word name "bed bug". The pests have been known by a variety of other informal names, including chilly billies, chinche bug, crimson rambler, heavy dragoon, mahogany flat, redcoat, and wall louse.

TICK

Ticks are small arachnids, part of the order Parasitiformes. Along with mites, they constitute the subclass Acari. Ticks are ectoparasites (external parasites), living by feeding on the blood of mammals, birds, and sometimes reptiles and amphibians. Ticks had evolved by the Cretaceous period, the most common form of fossilisation being immersed in amber. Ticks are widely distributed around the world, especially in warm, humid climates.

Almost all ticks belong to one of two major families, the Ixodidae or hard ticks, which are difficult to crush, and the Argasidae or soft ticks. Adults have ovoid or pear-shaped bodies which become engorged with blood when they feed, and eight legs. As well as having a hard shield on their dorsal surfaces, hard ticks have a beak-like structure at the front containing the mouthparts whereas soft ticks have their mouthparts on the underside of the body. Both families locate a potential host by odour or from changes in the environment.

Ticks have four stages to their lifecycle, namely egg, larva, nymph, and adult. Ixodid ticks have three hosts, taking at least a year to complete their lifecycle. Argasid ticks have up to seven nymphal stages (instars), each one requiring a blood meal. Because of their habit of ingesting blood, ticks are vectors of at least twelve diseases that affect humans and other animals.

Biology

Taxonomy and Phylogeny

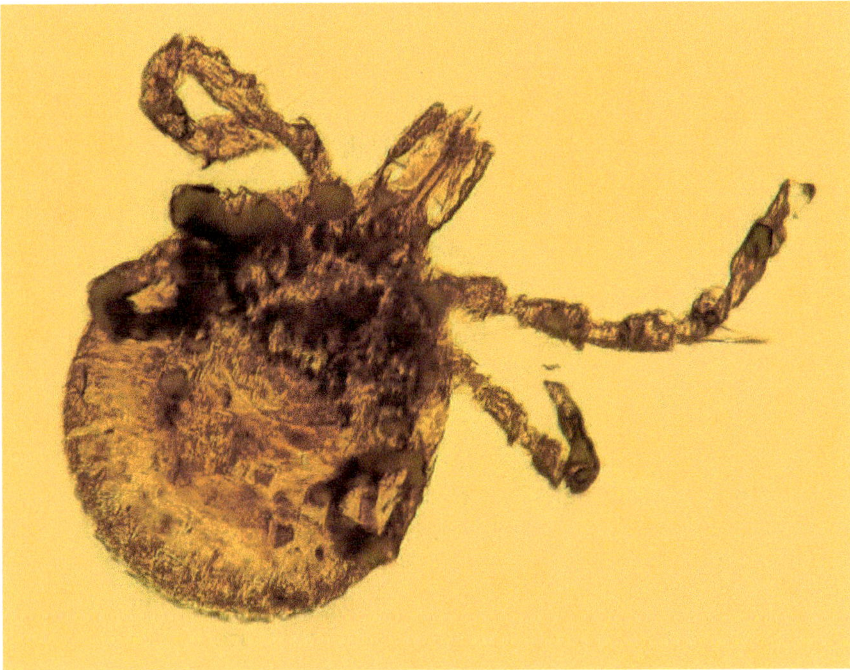

Fossilised tick in Dominican amber.

Fossilized ticks are known from the Cretaceous onwards, most commonly in amber. They most likely originated in the Cretaceous (65 to 146 million years ago), with most of the evolution and dispersal occurring during the Tertiary (5 to 65 million years ago). The oldest example is an argasid bird tick from Cretaceous New Jersey amber. The younger Baltic and Dominican ambers have also yielded examples which can be placed in living genera.

There are three families of ticks. The two large ones are the sister families of Ixodidae (hard ticks) and Argasidae (soft ticks). The third is Nuttalliellidae, named for the bacteriologist George Nuttall. It comprises a single species, *Nuttalliella namaqua*, and is the most basal lineage. Ticks are closely related to the mites, within the subclass Acarina. rDNA analysis suggests that the Ixodidae are a clade, but that the Argasidae may be paraphyletic.

The Ixodidae contains over 700 species of hard ticks with a scutum or hard shield, which the Argasidae lack. The Argasidae contains about 200 species; the genera accepted as of 2010 are *Antricola*, *Argas*, *Nothoaspis*, *Ornithodoros* and *Otobius*.They have no scutum, and the capitulum (mouth and feeding parts) is concealed beneath the body. The family Nuttalliellidae contains only a single species, *Nuttalliella namaqua*, a tick found in southern Africa from Tanzania to Namibia and South Africa.

The phylogeny of the Ixodida within the Acari is shown in the cladogram, based on a 2014 maximum parsimony study of amino acid sequences of twelve mitochondrial proteins. The Argasidae appear monophyletic in this study.

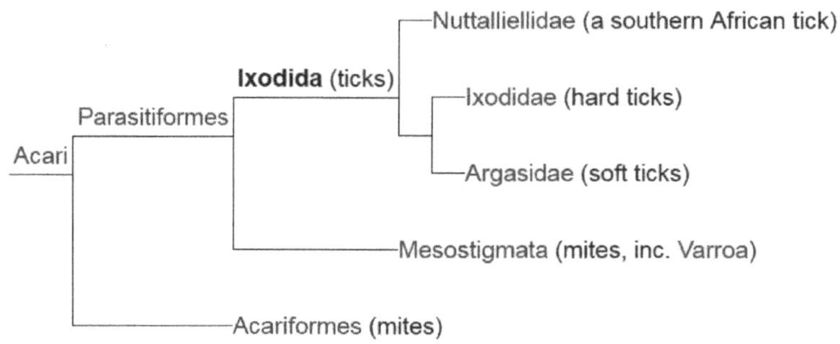

Range and Habitat

Tick species are widely distributed around the world, but they tend to flourish more in countries with warm, humid climates, because they require a certain amount of moisture in the air to undergo metamorphosis, and because low temperatures inhibit their development from egg to larva. Ticks are also widely distributed among host taxa, which include marsupial and placental mammals, birds, reptiles such as snakes, iguanas and lizards, and amphibians. Ticks of domestic animals cause considerable harm to livestock by transmission of many species of pathogen, as well as causing anaemia and damaging wool and hides. Some of the most debilitating species occur in tropical countries. Tropical bont ticks affect most domestic animals and occur in Africa and the Caribbean. The spinose ear tick has a worldwide distribution, the young feeding inside the ears of cattle and wild animals.

In general, ticks are to be found wherever their host species occur. Migrating birds carry ticks with them on their journeys; a study of migratory birds passing through Egypt found more than half the bird species examined were carrying ticks. The species of tick often differed between the autumn and spring migrations, probably because of the seasonal periodicities of the different species.

For an ecosystem to support ticks, it must satisfy two requirements: the population density of host species in the area must be high enough, and humidity must be high enough for ticks to remain hydrated. Due to their role in transmitting Lyme disease, ixodid ticks, particularly *I. scapularis*, have been studied using geographic information systems to develop predictive models for ideal tick habitats. According to these studies, certain features of a given microclimate – such as sandy soil, hardwood trees, rivers, and the presence of deer – were determined to be good predictors of dense tick populations.

Anatomy and Physiology

Ticks, like mites, are arthropods that have lost the segmentation of the abdomen that their ancestors had, and there has subsequently been a fusion of the abdomen with the cephalothorax. The tagmata typical of other Chelicera have been replaced by two new body sections, the anterior *capitulum* (or *gnathosoma*), which is retractable and contains the mouthparts, and the posterior *idiosoma* which contains the legs, digestive tract, and reproductive organs. The capitulum is a feeding structure with mouthparts adapted for piercing skin and sucking blood; it is only the front of the head and contains neither the brain nor the eyes. The ventral side of the idiosoma bears sclerites, and the gonopore is located between the fourth pair of legs. In the absence of segmentation, the positioning of the eyes, limbs and gonopore on the idiosoma provide the only locational guidance.

A hard-bodied tick of the family Ixodidae, the lone star tick.

Larval ticks hatch with six legs, acquiring the other two after a blood meal and molting into the nymph stage. In the nymphal and adult stages, ticks have eight legs, each of which has seven segments and is tipped with a pair of claws. The legs are sometimes ornamented and usually bear sensory or tactile hairs. In addition to being used for locomotion, the tarsus of leg I contains a unique sensory structure, Haller's organ, which can detect odors and chemicals emanating from the host, as well as sensing changes in temperature and air currents. When not being used for walking, the legs remain tightly folded against the body.

In nymphs and adults of the family Ixodidae, the *capitulum* is prominent and projects forwards from the body, a feature not present in the Argasidae. The eyes are close to the sides of the scutum, and the large spiracles are located just behind the coxae of the fourth pair of legs. The hard, protective scutellum, characteristic of this family, covers the whole dorsal surface in males but is restricted to a small, shield-like structure behind the capitulum in females and nymphs. They differ, too, in their life cycle; Ixodidae that attach to a host bite painlessly and generally unnoticed, and they remain in place until they engorge and are ready to change their skin; this process may take days or weeks. Some species drop off the host to moult in a safe place, whereas others remain on the same host and only drop off once they are ready to lay their eggs.

A soft-bodied tick of the family Argasidae, beside eggs it has just laid.

The body of the soft tick, family Argasidae, is pear-shaped or oval with a rounded anterior portion. The mouthparts cannot be seen from above as they are on the ventral surface. The cuticle is leathery; there is often a centrally positioned dorsal plate, with ridges which project slightly above the surrounding surface, but there is no decoration. A pattern of small, circular depressed areas show where muscles are attached to the interior of the integument. The eyes are on the sides of the body, the spiracles open between legs 3 and 4, and males and females only differ in the structure of the genital pore.

The Nuttalliellidae can be distinguished from both ixodid and argasid ticks by a combination of a projecting capitulum at the front and a soft, leathery skin. Other distinguish characteristics include the position of the stigmata, the lack of setae, the strongly corrugated integument, and the form of the fenestrated plates.

Diet and Feeding

Ticks satisfy all of their nutritional requirements as ectoparasites, feeding on a diet of blood. They are obligate hematophages, needing blood to survive and move from one stage of life to another. Ticks can fast for long periods but eventually die if unable to find a host. This behavior evolved approximately 120 million years ago through adaptation to blood-feeding. The behavior evolved independently in the separate tick families, with differing host-tick interactions driving the evolutionary change.

Some ticks attach quickly while others wander around looking for thinner skin such as is found on the ears of mammals. Depending on the species and the life stage, preparing to feed can take from ten minutes to two hours. On locating a suitable feeding spot, the tick grasps the host's skin and cuts into the surface. They extract blood by cutting a hole in the host's epidermis, into which they insert their hypostome, and keep the blood from clotting by excreting an anticoagulant or platelet aggregation inhibitor.

A questing tick, fingers for scale.

Ticks find their hosts by detecting animals' breath and body odors, or by sensing body heat, moisture and vibrations. They are incapable of flying or jumping, but many tick species, particularly Ixodidae, lie in wait in a position known as "questing". While questing, ticks cling to leaves and grasses by their third and fourth pair of legs. They hold the first pair of legs outstretched, waiting to grasp and climb on to any passing host. Tick questing heights tend to be correlated with the size of the desired host; nymphs and small species tend to quest close to the ground where they may encounter small mammalian or bird hosts; adults climb higher into the vegetation where larger hosts may be encountered. Some species are hunters and lurk near places where hosts may rest. On receiving an olfactory or other stimulus, they crawl or run across the intervening surface.

Other ticks, mainly the Argasidae, are "nidicolous", finding hosts in their nests or burrows, and in caves in the case of bats. They use the same stimuli as non-nidicolous species to identify hosts, with body heat and odors often being the main factors. Many of them feed primarily on birds, though some *Ornithodoros*, for example, feed on mammals. Both groups feed rapidly, typically biting painfully and drinking their fill within minutes. None of the species stick to the host in the way that hard ticks do. Unlike the Ixodidae that have no fixed dwelling place except on the host, they live in sand or in crevices near animal dens or nests, or in human dwellings where they come out nightly to attack roosting birds, or emerge when they detect carbon dioxide in the breath of their hosts.

In the Ixodidae, the tick stays in place until it is completely engorged. Its weight may increase by 200 to 600 times as compared to its weight before it started feeding. To accommodate this large expansion, cell division takes place and its cuticle grows larger;

the tick may remain attached for days or weeks, depending on species, life stage and host. In the Argasidae, the tick's cuticle stretches to accommodate the fluid ingested but does not grow new cells, the weight of the tick increasing five to tenfold over the unfed state. The tick then drops off the host and typically remains in the nest or burrow until its host returns to provide its next meal.

Ecology

Mites and nematodes feed on ticks, which are also a minor nutritional resource for birds. More importantly, they carry diseases as the primary hosts of pathogens such as spirochaetes, and without their agency, the organisms could not reach their secondary hosts. The diseases caused may debilitate their victims, and ticks may thus be assisting in controlling animal populations and preventing overgrazing.

Certain infectious diseases of humans and other animals can be transmitted by ticks, the species of tick involved tending to be those with a wide host range. Spread of disease in this way is enhanced by the extended time during which a tick remains attached, during which time the mobile host can be carried long distances, or in the case of bird hosts, across the sea. The infective agents can be present not only in the adult tick but also in the eggs produced plentifully by the females. Many tick species have extended their ranges as a result of the movements of people, their pets and livestock. With increasing participation in outdoor activities such as wilderness hikes, more people and their dogs may find themselves exposed to attack.

Life Cycle

Both ixodid and argasid ticks have four lifecycle stages: egg, larva, nymph, and adult.

Ixodidae

Ixodid ticks require three hosts, and their life cycle takes at least one year to complete. Up to 3,000 eggs are laid on the ground by an adult female tick. When the larvae emerge, they feed primarily on small mammals and birds. After feeding, they detach from their host and molt to nymphs on the ground, which then feed on larger hosts and molt to adults. Female adults attach to larger hosts, feed, and lay eggs, while males feed very little and occupy larger hosts primarily for mating.

Argasidae

Argasid ticks, unlike ixodid ticks, may go through up to seven nymphal stages (instars), requiring a meal of blood each time. Their life cycles range from months to years. The adult female argasid tick can lay a few hundred to over a thousand eggs over the course of her lifetime. Larvae feed very quickly and detach to molt into nymphs. Both male and female adults feed on blood, and they mate off the host. During feeding, any excess fluid is excreted by the coxal glands, a process which is unique to argasid ticks.

Relationship with Humans

Tick-Borne Disease

A sign in a Lithuanian forest warning of high risk of tick-borne encephalitis infection.

Ticks are implicated in the transmission of a number of infections caused by pathogens such as bacteria, viruses and protozoa. Sometimes the tick harbours more than one type of pathogen, making diagnosis of the infection more difficult. Species of the bacterium *Rickettsia* are responsible for typhus, rickettsialpox, Boutonneuse fever, African tick bite fever, Rocky Mountain spotted fever, Flinders Island spotted fever and Queensland tick typhus (Australian tick typhus). Other tick-borne diseases include Lyme disease and Q fever, Colorado tick fever, Crimean Congo hemorrhagic fever, tularemia, tick-borne relapsing fever, babesiosis, ehrlichiosis, and tick-borne meningoencephalitis, as well as bovine anaplasmosis and probably the Heartland virus.

Some species, notably the Australian paralysis tick, are also intrinsically venomous and can cause tick paralysis. Eggs can become infected with pathogens inside a female tick's ovaries, in which case the larval ticks are infectious immediately at hatching, before feeding on their first host. Tropical bont ticks transmit the rickettsial disease heartwater which can be particularly devastating in cattle. The ticks carried by migratory birds may act as reservoirs and vectors of infectious diseases. Over twenty strains of pathogenic virus were found in the autumn in the Egyptian migratory bird study.

Not all ticks in an infective area are infected with pathogens, and both attachment of the tick and a long-feeding session seem to be necessary for transmission of these diseases to take place. Thus tick bites often do not lead to infection, especially if the ticks are removed within 36 hours. Adult ticks can be removed with fine-tipped tweezers or proprietary tick removal tools, disinfecting the wound. It is also possible to freeze them off with a medical wart remover. If the tick's head and mouthparts break off during removal, they can be removed with tweezers like a splinter.

Population Control Measures

Researcher collecting ticks using the "tick dragging" method.

With the possible exception of widespread DDT use in the Soviet Union, attempts to limit the population or distribution of disease-causing ticks have been quite unsuccessful. The parasitoid chalcid wasp *Ixodiphagus hookeri* has been investigated for its potential to control tick populations. It lays its eggs into ticks; the hatching wasps kill their hosts. Another natural form of control for ticks is the guineafowl, a bird species which consumes mass quantities of ticks.

Topical tick medicines may be toxic to animals and humans. The synthetic pyrethroid insecticide Phenothrin in combination with the hormone analogue methoprene was a popular topical flea and tick therapy for felines. Phenothrin kills adult ticks, while methoprene kills eggs. However, some products have been withdrawn, while others are known to cause adverse reactions.

SCABIES

Scabies, previously known as the seven-year itch, is a contagious skin infestation by the mite *Sarcoptes scabiei*.The most common symptoms are severe itchiness and a pimple-like rash. Occasionally tiny burrows may be seen in the skin. When first infected, usually two to six weeks are required before symptoms occur. If a person develops a second infection later in life, symptoms may begin within a day. These symptoms can be present across most of the body or just certain areas such as the wrists, between fingers, or along the waistline. The head may be affected, but this is typically only in young children. The itch is often worse at night. Scratching may cause skin breakdown and an additional bacterial infection of the skin.

Scabies is caused by infection with the female mite *Sarcoptes scabiei*. The mites burrow into the skin to live and deposit eggs. The symptoms of scabies are due to an allergic reaction to the mites. Often only between ten and fifteen mites are involved in an infection. Scabies is most often spread during a relatively long period of direct skin contact with an infected person such as that which may occur during sex. Spread of disease may occur even if the person has not developed symptoms yet. Crowded living conditions such as those found in child care facilities, group homes, and prisons increase the risk of spread. Areas with a lack of access to water also have higher rates of disease. **Crusted scabies** is a more severe form of the disease. It typically only occurs in those with a poor immune system and people may have millions of mites, making them much more contagious. In these cases spread of infection may occur during brief contact or via contaminated objects. The mite is very small and usually not directly visible. Diagnosis is based on the signs and symptoms.

A number of medications are available to treat those infected, including permethrin, crotamiton and lindane creams and ivermectin pills. Sexual contacts within the last month and people who live in the same house should also be treated at the same time. Bedding and clothing used in the last three days should be washed in hot water and dried in a hot dryer. As the mite does not live for more than three days away from human skin more washing is not needed. Symptoms may continue for two to four weeks following treatment. If after this time there continue to be symptoms retreatment may be needed.

Scabies is one of the three most common skin disorders in children, along with ringworm and bacterial skin infections. As of 2010 it affects approximately 100 million people (1.5% of the world population) and is equally common in both sexes. The young and the old are more commonly affected. It also occurs more commonly in the developing world and tropical climates. The word scabies is from Latin: *scabere*, "to scratch". Other animals do not spread human scabies.Infection in other animals is typically caused by slightly different but related mites and is known as sarcoptic mange.

Signs and Symptoms

The characteristic symptoms of a scabies infection include intense itching and superficial burrows. The burrow tracks are often linear, to the point that a neat "line" of four or more closely placed and equally developed mosquito-like "bites" is almost diagnostic of the disease. Because the host develops the symptoms as a reaction to the mites' presence over time, there is typically a delay of four to six weeks between the onset of infestation and the onset of itching. Similarly, symptoms often persist for one to several weeks after successful eradication of the mites. As noted, those re-exposed to scabies after successful treatment may exhibit symptoms of the new infestation in a much shorter period—as little as one to four days.

Itching

In the classic scenario, the itch is made worse by warmth, and is usually experienced as being worse at night, possibly because there are fewer distractions. As a symptom, it is less common in the elderly.

Rash

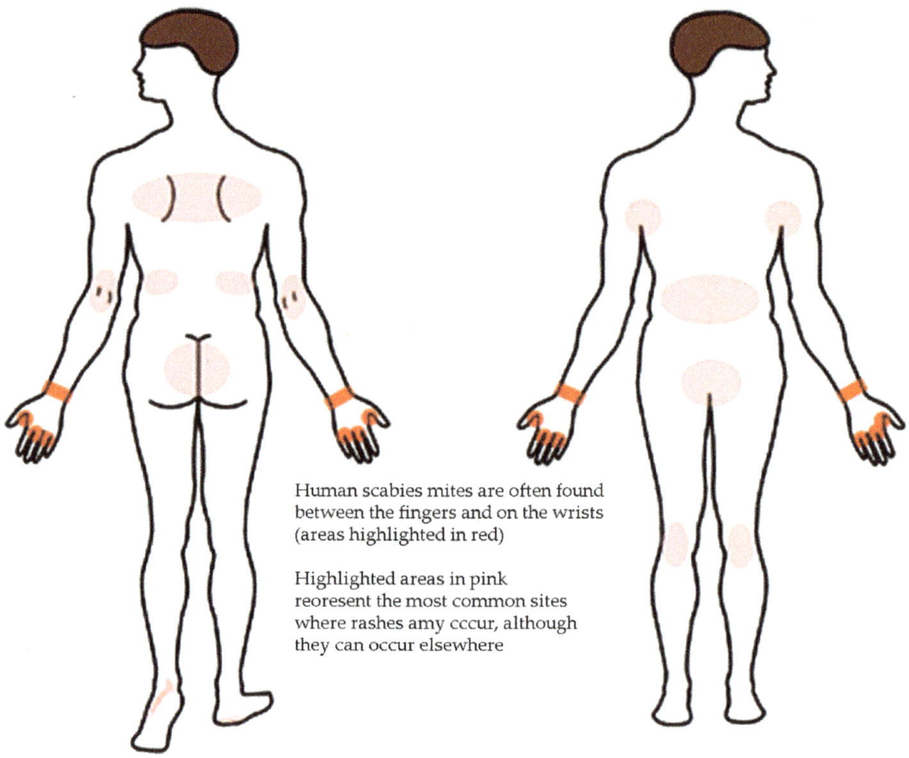

Commonly involved sites of rashes of scabies

The superficial burrows of scabies usually occur in the area of the finger webs, feet, ventral wrists, elbows, back, buttocks, and external genitals. Except in infants and the immunosuppressed, infection generally does not occur in the skin of the face or scalp. The burrows are created by excavation of the adult mite in the epidermis.

In most people, the trails of the burrowing mites are linear or *s*-shaped tracks in the skin often accompanied by rows of small, pimple-like mosquito or insect bites. These signs are often found in crevices of the body, such as on the webs of fingers and toes, around the genital area, in stomach folds of the skin, and under the breasts of women.

Symptoms typically appear two to six weeks after infestation for individuals never before exposed to scabies. For those having been previously exposed, the symptoms can appear within several days after infestation. However, it is not unknown for symptoms to appear after several months or years. Acropustulosis, or blisters and pustules on the palms and soles of the feet, are characteristic symptoms of scabies in infants.

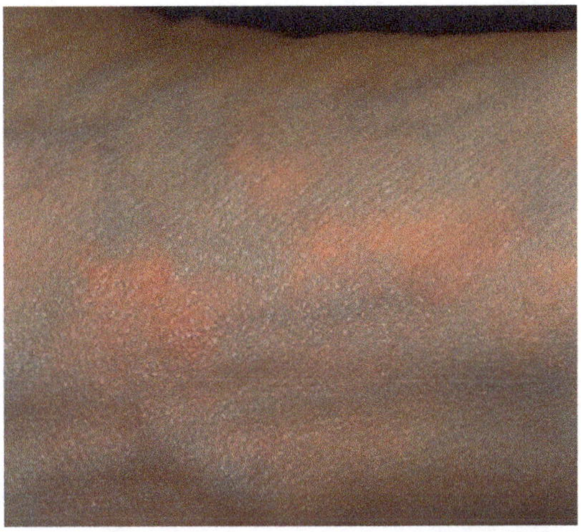

Scabies of the foot

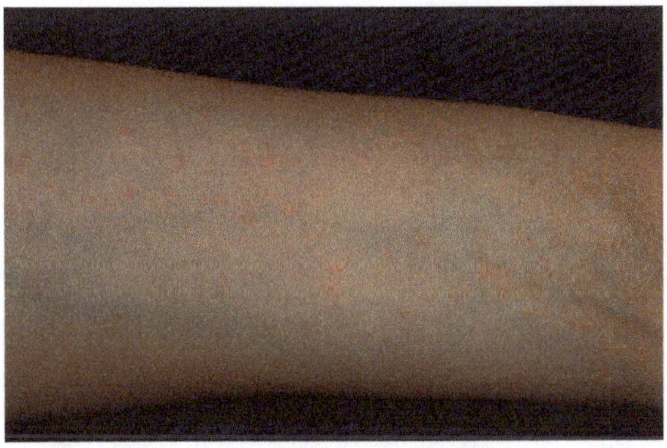

Scabies of the arm

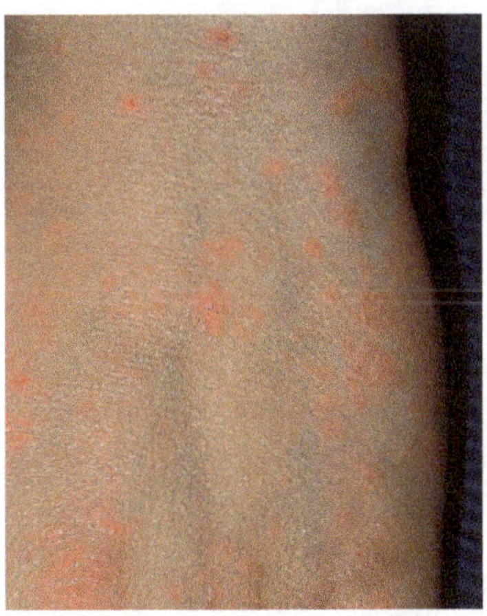

Scabies of the hand

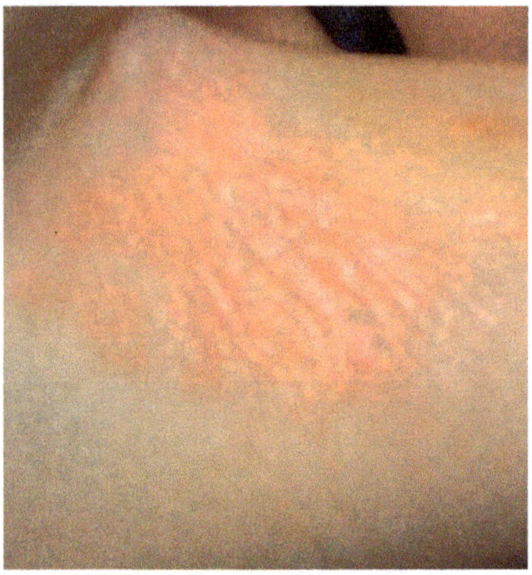

Scabies of the finger

Crusted scabies

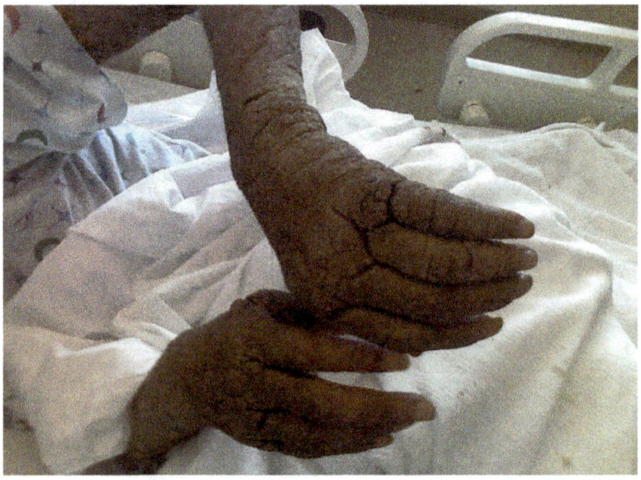

Crusted scabies in a person with AIDS.

The elderly and people with an impaired immune system, such as HIV, cancer, or those on immunosuppressive medications, are susceptible to crusted scabies (formerly called Norwegian scabies). On those with weaker immune systems, the host becomes a more fertile breeding ground for the mites, which spread over the host's body, except the face. Sufferers of crusted scabies exhibit scaly rashes, slight itching, and thick crusts of skin that contain thousands of mites. Such areas make eradication of mites particularly difficult, as the crusts protect the mites from topical miticides/scabicides, necessitating prolonged treatment of these areas.

Cause

Scabies Mite

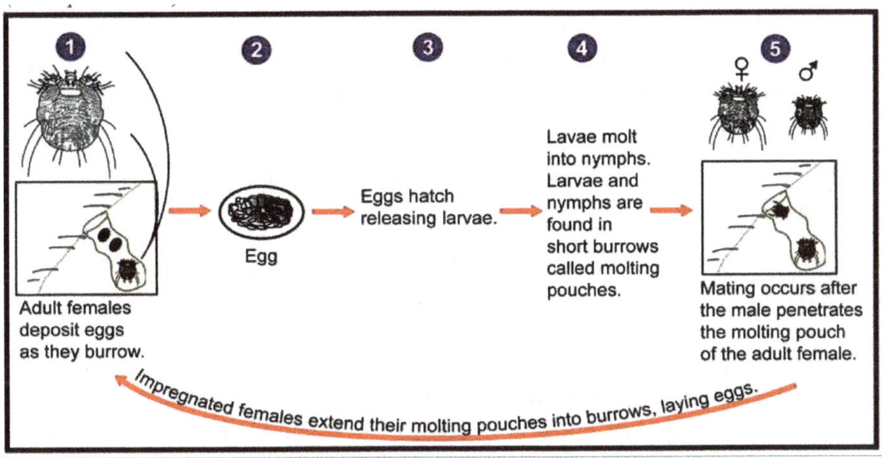

Life cycle of scabies

In the 18th century, Italian biologist Diacinto Cestoni (1637–1718) described the mite now called *Sarcoptes scabiei*, variety *hominis*, as the cause of scabies. *Sarcoptes* is a genus of skin parasites and part of the larger family of mites collectively known as scab mites. These organisms have eight legs as adults, and are placed in the same phylogenetic class (Arachnida) as spiders and ticks.

Sarcoptes scabiei mites are under 0.5 mm in size but are sometimes visible as pinpoints of white. Pregnant females tunnel into the dead, outermost layer (stratum corneum) of a host's skin and deposit eggs in the shallow burrows. The eggs hatch into larvae in three to ten days. These young mites move about on the skin and molt into a "nymphal" stage, before maturing as adults, which live three to four weeks in the host's skin. Males roam on top of the skin, occasionally burrowing into the skin. In general, the total number of adult mites infesting a healthy hygienic person with non-crusted scabies is small; about 11 females in burrows, on average.

The movement of mites within and on the skin produces an intense itch, which has the characteristics of a delayed cell-mediated inflammatory response to allergens. IgE antibodies are present in the serum and the site of infection, which react to multiple protein allergens in the body of the mite. Some of these cross-react to allergens from house dust mites. Immediate antibody-mediated allergic reactions (wheals) have been elicited in infected persons, but not in healthy persons; immediate hypersensitivity of this type is thought to explain the observed far more rapid allergic skin response to reinfection seen in persons having been previously infected (especially having been infected within the previous year or two).

Transmission

Scabies is contagious and can be contracted through prolonged physical contact with an infested person. This includes sexual intercourse, although a majority of cases are acquired through other forms of skin-to-skin contact. Less commonly, scabies infestation can happen through the sharing of clothes, towels, and bedding, but this is not a major mode of transmission; individual mites can only survive for two to three days, at most, away from human skin. As with lice, a latex condom is ineffective against scabies transmission during intercourse, because mites typically migrate from one individual to the next at sites other than the sex organs.

Healthcare workers are at risk of contracting scabies from patients, because they may be in extended contact with them.

Pathophysiology

The symptoms are caused by an allergic reaction of the host's body to mite proteins, though exactly which proteins remains a topic of study. The mite proteins are also present from the gut, in mite feces, which are deposited under the skin. The allergic reaction is both of the delayed (cell-mediated) and immediate (antibody-mediated) type, and involves IgE (antibodies, it is presumed, mediate the very rapid symptoms on reinfection). The allergy-type symptoms (itching) continue for some days, and even several weeks, after all mites are killed. New lesions may appear for a few days after mites are eradicated. Nodular lesions from scabies may continue to be symptomatic for weeks after the mites have been killed.

A 2016 study found rates of scabies was negatively related to temperature and positively related with humidity.

Diagnosis

Scabies may be diagnosed clinically in geographical areas where it is common when diffuse itching presents along with either lesions in two typical spots or there is itchiness of another household member.The classical sign of scabies is the burrows made by the mites within the skin. To detect the burrow, the suspected area is rubbed with ink from a fountain pen or a topical tetracycline solution, which glows under a special light. The skin is then wiped with an alcohol pad. If the person is infected with scabies, the characteristic zigzag or *S* pattern of the burrow will appear across the skin; however, interpreting this test may be difficult as the burrows are scarce and may be obscured by scratch marks. A definitive diagnosis is made by finding either the scabies mites or their eggs and fecal pellets. Searches for these signs involve either scraping a suspected area, mounting the sample in potassium hydroxide and examining it under a microscope, or using dermoscopy to examine the skin directly.

Magnified view of a burrowing trail of the scabies mite: The scaly patch on the left was caused by the scratching and marks the mite's entry point into the skin. The mite has burrowed to the top-right, where it can be seen as a dark spot at the end.

Differential Diagnosis

Symptoms of early scabies infestation mirror other skin diseases, including dermatitis, syphilis, erythema multiforme, various urticaria-related syndromes, allergic reactions, and other ectoparasites such as lice and fleas.

Prevention

Mass treatment programs that use topical permethrin or oral ivermectin have been effective in reducing the prevalence of scabies in a number of populations. No vaccine is available for scabies. The simultaneous treatment of all close contacts is recommended, even if they show no symptoms of infection (asymptomatic), to reduce rates of recurrence. Since mites can survive for only two to three days without a host, other objects in the environment pose little risk of transmission except in the case of crusted scabies, thus cleaning is of little importance. Rooms used by those with crusted scabies require thorough cleaning.

Management

A number of medications are effective in treating scabies. Treatment should involve the entire household, and any others who have had recent, prolonged contact with the infested individual. Options to control itchiness include antihistamines and prescription anti-inflammatory agents. Bedding, clothing and towels used during the previous three days should be washed in hot water and dried in a hot dryer.

Permethrin

Permethrin is the most effective treatment for scabies, and remains the treatment of choice. It is applied from the neck down, usually before bedtime, and left on for about eight to 14 hours, then washed off in the morning. Care should be taken to coat the entire skin surface, not just symptomatic areas; any patch of skin left untreated can provide a "safe haven" for one or more mites to survive. One application is normally sufficient, as permethrin kills eggs and hatchlings as well as adult mites, though many physicians recommend a second application three to seven days later as a precaution. Crusted scabies may require multiple applications, or supplemental treatment with oral ivermectin (below). Permethrin may cause slight irritation of the skin that is usually tolerable.

Ivermectin

Oral Ivermectin is effective in eradicating scabies, often in a single dose. It is the treatment of choice for crusted scabies, and is sometimes prescribed in combination with a topical agent. It has not been tested on infants, and is not recommended for children under six years of age.

Topical ivermectin preparations have been shown to be effective for scabies in adults, though only one such formulation is available in the United States at present, and it is not FDA approved as a scabies treatment. It has also been useful for sarcoptic mange (the veterinary analog of human scabies).

Others

Other treatments include lindane, benzyl benzoate, crotamiton, malathion, and sulfur preparations. Lindane is effective, but concerns over potential neurotoxicity has limited its availability in many countries. It is banned in California, but may be used in other states as a second-line treatment. Sulfur ointments or benzyl benzoate are often used in the developing world due to their low cost; 10% sulfur solutions have been shown to be effective, and sulfur ointments are typically used for at least a week, though many people find the odor of sulfur products unpleasant. Crotamiton has been found to be less effective than permethrin in limited studies. Crotamiton or sulfur preparations are sometimes recommended instead of permethrin for children, due to concerns over dermal absorption of permethrin.

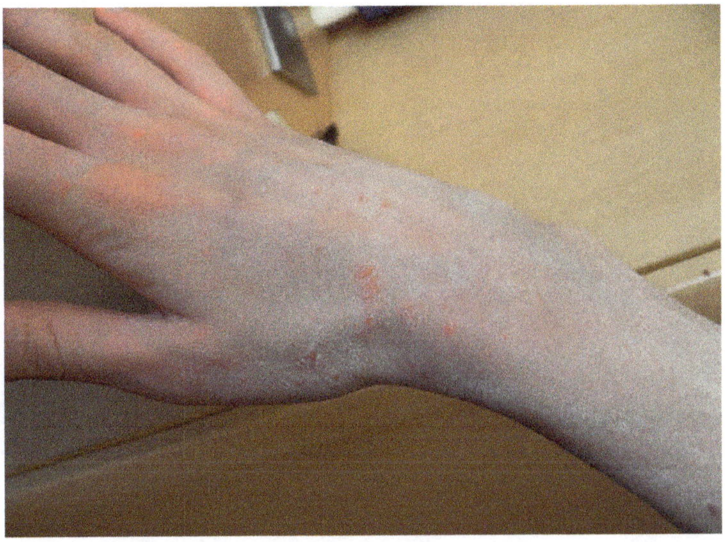

Day 4

Day 8 (treatment begins)

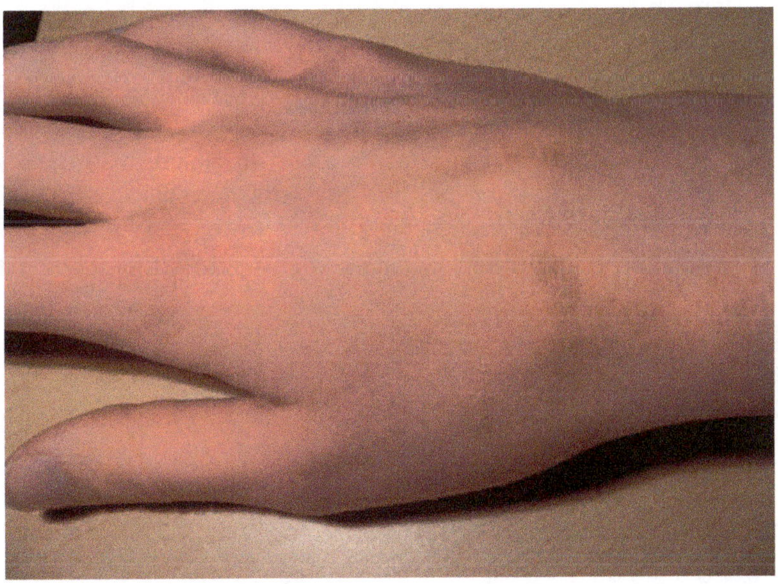

Day 12 (under treatment)

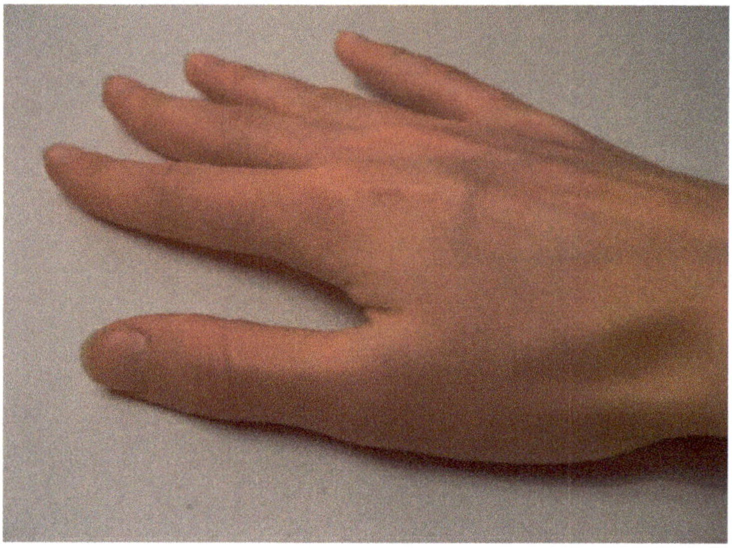

Healed

Communities

Scabies is endemic in many developing countries, where it tends to be particularly problematic in rural and remote areas. In such settings community wide control strategies are required to reduce the rate of disease, as treatment of only individuals is ineffective due to the high rate of reinfection. Large-scale mass drug administration strategies may be required where coordinated interventions aim to treat whole communities in one concerted effort. Although such strategies have shown to be able to reduce the burden of scabies in these kinds of communities, debate remains about the best strategy to adopt, including the choice of drug.

The resources required to implement such large-scale interventions in a cost-effective and sustainable way are significant. Furthermore, since endemic scabies is largely restricted to poor and remote areas, it is a public health issue that has not attracted much attention from policy makers and international donors.

Epidemiology

Scabies is one of the three most common skin disorders in children, along with tinea and pyoderma. As of 2010 it affects approximately 100 million people (1.5% of the population) and is equally common in both genders. The mites are distributed around the world and equally infect all ages, races, and socioeconomic classes in different climates. Scabies is more often seen in crowded areas with unhygienic living conditions. Globally as of 2009, an estimated 300 million cases of scabies occur each year, although various parties claim the figure is either over- or underestimated. About 1–10% of the global population is estimated to be infected with scabies, but in certain populations, the infection rate may be as high as 50–80%.

Wax figurine of a man with Norwegian scabies.

Scabies has been observed in humans since ancient times. Archeological evidence from Egypt and the Middle East suggests scabies was present as early as 494 BC. The first recorded reference to scabies is believed to be from the Bible – it may be a type of "leprosy" mentioned in Leviticus *circa* 1200 BC or be mentioned among the curses of Deuteronomy 28. In the fourth century BC, Aristotle reported on "lice" that "escape from little pimples if they are pricked" – a description consistent with scabies.

The Roman encyclopedist and medical writer Aulus Cornelius Celsus (c. 25 BC – c. 50 AD) is credited with naming the disease "scabies" and describing its characteristic features. The parasitic etiology of scabies was documented by the Italian physician Giovanni Cosimo Bonomo (1663–1696) in his 1687 letter, "Observations concerning the fleshworms of the human body". Bonomo's description established scabies as one of the first human diseases with a well-understood cause.

In Europe in the late 19th through mid-20th centuries, a sulfur-bearing ointment called by the medical eponym of Wilkinson's ointment was widely used for topical treatment of scabies. The contents and origins of several versions of the ointment were detailed in correspondence published in the *British Medical Journal* in 1945.

Society and Culture

The International Alliance for the Control of Scabies (IACS) was started in 2012, and brings together over 70 researchers, clinicians and public health experts from more than 15 different countries. It has managed to bring the global health implications of scabies to the attention of the World Health Organization.Consequently, the WHO has included scabies on its official list of neglected tropical diseases and other neglected conditions.

Other Animals

Scabies may occur in a number of domestic and wild animals; the mites that cause these infestations are of different subspecies from the one typically causing the human form. These subspecies can infest animals that are not their usual hosts, but such infections do not last long. Scabies-infected animals suffer severe itching and secondary skin infections. They often lose weight and become frail.

The most frequently diagnosed form of scabies in domestic animals is sarcoptic mange, caused by the subspecies *Sarcoptes scabiei canis*, most commonly in dogs and cats. Sarcoptic mange is transmissible to humans who come into prolonged contact with infested animals,and is distinguished from human scabies by its distribution on skin surfaces covered by clothing. Scabies-infected domestic fowl suffer what is known as "scaly leg". Domestic animals that have gone feral and have no veterinary care are frequently afflicted with scabies and a host of other ailments. Nondomestic animals have also been observed to suffer from scabies. Gorillas, for instance, are known to be susceptible to infection via contact with items used by humans.

A street dog in Bali, Indonesia, suffers from sarcoptic mange.

Research

Moxidectin is being evaluated as a treatment for scabies. It is established in veterinary medicine to treat a range of parasites, including sarcoptic mange. Its advantage over ivermectin is its longer duration of action.

REFERENCES

1. "Oeciacus hirundinis (Lamarck, 1816)". Integrated Taxonomic Information System. 1996. Retrieved 15 September 2010.
2. Bat bugs are visually indistinguishable from bed bugs, and may also occasionally feed on human hosts.
3. Goddard, J; deShazo, R (1 April 2009). "Bed bugs (Cimex lectularius) and clinical consequences of their bites". JAMA: The Journal of the American Medical Association. 301 (13): 1358–66. doi:10.1001/jama.2009.405. PMID 19336711.
4. Reinhardt, Klaus; Siva-Jothy, Michael T. (Jan 2007). "Biology of the Bed Bugs (Cimicidae)" (PDF). Annual Review of Entomology. 52: 351–374. doi:10.1146/annurev.ento.52.040306.133913. PMID 16968204. Archived (PDF) from the original on 5 July 2010. Retrieved 26 May 2010.
5. "What Are Bed Bugs? How To Kill Bed Bugs". Medical News Today. MediLexicon International Ltd. 20 July 2009. Retrieved 27 May 2010.
6. Doggett SL, Russell R (November 2009). "Bed bugs – What the GP needs to know". Aust Fam Physician. 38 (11): 880–4. PMID 19893834.
7. Jerome Goddard; Richard deShazo (2009). "Bed bugs (Cimex lectularius) and

clinical consequences of their bites". Journal of the American Medical Association. 301 (13): 1358–1366. doi:10.1001/jama.2009.405. PMID 19336711.

8. Kilpenen, O.; Vagn Jensen, K-M.; Kristensen, M. "Bed Bug Problems in Denmark, with a European Perspective" (PDF): 395–399. Retrieved 27 May 2010.
9. Robinson, WH; Bajomi, D. (eds) (2008). Proceedings of the Sixth International Conference on Urban Pests. Hungary: OOK-Press Kft.
10. Kolb, Adam; Needham, Glen R.; Neyman, Kimberly M.; High, Whitney A. (2009). "Bedbugs". Dermatologic Therapy. 22 (4): 347–52. doi:10.1111/j.1529-8019.2009.01246.x. PMID 19580578.
11. James, William D.; Berger, Timothy G. (2006). Andrews' Diseases of the Skin: clinical Dermatology. Saunders Elsevier. ISBN 0-7216-2921-0.
12. Melnick, Meredith (12 May 2011). "Study: Bedbugs May Carry MRSA; Germ Transmission Unclear | TIME.com". Healthland.time.com. Retrieved 11 November 2013.
13. "Do Bedbugs Spread MRSA and VRE?". Webmd.com. 11 May 2011. Retrieved 11 November 2013.
14. Adelman, ZN (Aug 2013). "Bed bugs and infectious disease: a case for the arboviruses". PLoS Pathogens. 9 (8): e1003462. doi:10.1371/journal.ppat.1003462. PMC 3744395 . PMID 23966852.
15. Cannet, Arnaud; Akhoundi, Mohammad; Berenger, Jean-Michel; Michel, Gregory; Marty, Pierre; Delaunay, Pascal (2015). "A review of data on laboratory colonies of bed bugs (Cimicidae), an insect of emerging medical relevance". Parasite. 22: 21. doi:10.1051/parasite/2015021. ISSN 1776-1042. PMID 26091944.
16. Quarles, William (March 2007). "Bed Bugs Bounce Back" (PDF). IPM Practitioner. Berkeley, CA: Bio-Integral Resource Center. 24 (3/4): 1–8. Retrieved 27 May 2010.
17. Olson, Joelle; Eaton, Marc; Kells, Stephen; Morin, Victor; Wang, Changlu (2013). "Cold Tolerance of Bed Bugs and Practical Recommendations for Control". Journal of Economic Entomology. 106 (6): 2433–2441. doi:10.1603/EC13032. PMID 24498745. Retrieved 9 July 2014.
18. Benoit, J. B.; del Grosso, N.; Yoder, J. A.; Denlinger, D. L. (2007). "Resistance to dehydration between bouts of blood feeding in the bed bug, Cimex lectularius, is enhanced by water conservation, aggregation, and quiescence". American Journal of Tropical Medicine and Hygiene. 76 (5): 987–993. ISSN 0002-9637. PMID 17488928. Retrieved 27 May 2010.
19. Herrmann, J.; Adler, C.; Hoffmann, G.; Reichmuth, C. (1999). "Efficacy of controlled atmospheres on Cimex lectularius (L.) (Heteroptera: Cimicidae) and Argas reflexus Fab. (Acari: Argasidae)" (PDF). In Robinson, Wm H.; Rettich, F.; Rambo, G.W. Proceedings of the 3rd International Conference on Urban Pests. Hronov, Czech Republic: Grafické Závody. p. 637. Retrieved 31 May 2010. (abstracted from a poster presentation in Prague, 19–22 Jul)
20. Storey, Malcom. "CIMICIDAE (bed bugs)". BioImages: The Virtual Field-

Guide (UK). bioimages.org.uk. Archived from the original on 14 August 2010. Retrieved 27 May 2010.

21. Mullen, Gary R.; Durden, Lance A. (8 May 2009). Medical and Veterinary Entomology, Second Edition. Academic Press. p. 80. ISBN 0-12-372500-3.
22. "Family CIMICIDAE". Australian Biological Resources Study: Australian Faunal Directory. Department of the Environment, Water, Heritage and the Arts (Australia). 2008. Retrieved 27 May 2010.
23. Richards, Luck (1999). Integument of Arthropods. University of Minnesota. p. 298. ISBN 978-0-8166-0073-1. Retrieved 1 December 2016.
24. Anderson, J. F.; Ferrandino, F. J.; McKnight, S.; Nolen, J.; Miller, J. (2009). "A carbon dioxide, heat and chemical lure trap for the bed bug, Cimex lectularius" (PDF). Medical and Veterinary Entomology. 23 (2): 99–105. doi:10.1111/j.1365-2915.2008.00790.x. Retrieved 27 May 2010.
25. Narinderpal Singh; Changlu Wang; Richard Cooper; Chaofeng Liu. "Interactions among Carbon Dioxide, Heat, and Chemical Lures in Attracting the Bed Bug, Cimex lectularius L. (Hemiptera: Cimicidae)". Psyche. 2012: 1–9. doi:10.1155/2012/273613.
26. Changlu Wang; Timothy Gibb; Gary W Bennett; Susan McKnight (2009). "Bed bug (Heteroptera: Cimicidae) attraction to pitfall traps baited with carbon dioxide, heat, and chemical lure.". Journal of Economic Entomology (8). 102(4):1580-5.
27. Robert L. Usinger (1966). Monograph of Cimicidae (Hemiptera – Heteroptera) (PDF). V11. The Thomas Say Foundation. p. 13.
28. Polanco, Andrea (2011). "Survivorship During Starvation for Cimex lectularius L.". Insects. 2: 232–242. doi:10.3390/insects2020232.
29. Milius, Susan (16 January 2010). "Do-it-yourself bed-bug detector". Science News. Society for Science and the Public. 177 (2): 88. Retrieved 27 May 2010. citingPolanco-Pinzón, Andrea; et al. (2009). "Survivorship and growth potential of modern bed bug populations (Cimex lectularius) in the United States". Conference proceedings of 57th Annual Meeting. Entomological Society of America.
30. A. L. Szalanski; J. W. Austin; J. A. McKern; C. D. Steelman; D. Miller (2006). "Time course analysis of bed bug, Cimex lectularius L., (Hemiptera: Cimicidae) blood meals using PCR". Journal of Agricultural and Urban Entomology. 23: 237–241.
31. A. L. Szalanski; J. W. Austin; J. A. McKern; D. Miller; C. D. Steelman; R. E. Gold (2006). "Isolation and characterization of human DNA from the bed bug Cimex lectularius L. (Heteroptera: Cimicidae)" (PDF). Journal of Agricultural and Urban Entomology. 23: 189–194.
32. I. Thomas; G. G. Kihiczak; R. A. Schwartz (2004). "Bed bug bites: a review". International Journal of Dermatology. 43 (6): 430–433. doi:10.1111/j.1365-4632.2004.02115.x. PMID 15186224.
33. Carayon, J. 1959 Insémination par "spermalège" et cordon conducteur de spermatozoids chez Stricticimex brevispinosus Usinger (Heteroptera, Cimicidae).

Rev. Zool. Bot. Afr. 60, 81–104.

34. Morrow E. H.; Arnqvist G. (2003). "Costly traumatic insemination and a female counter-adaptation in bed bugs". Proceedings of the Royal Society B. 270 (1531): 2377–2381. doi:10.1098/rspb.2003.2514. PMC 1691516 . PMID 14667354.
35. b Reinhardt K.; Naylor R.; Siva-Jothy M. T. (2003). "Reducing a cost of traumatic insemination: female bedbugs evolve a unique organ". Proceedings of the Royal Society B. 270 (1531): 2371–2375. doi:10.1098/rspb.2003.2515. PMC 1691512 . PMID 14667353.
36. Ryne, C. In press. "Homosexual interactions in bed bugs: Alarm pheromones as male recognition signals." Animal Behaviour.doi:10.1016/j.anbehav.2009.09.033
37. "This Bedbug's Life", The New York Times, 7 August 2010.
38. Harari A.; Brockman H. J.; Landholt P. J. (2000). "Intrasexual mounting in the beetle Diaprepes abbreviatus (L.)". Proceedings of the Royal Society B. 267 (1457): 2071–2079. doi:10.1098/rspb.2000.1251. PMC 1690776 . PMID 11416911.
39. "Scent of alarm identifies male bed bugs". Science News. 29 October 2009. Retrieved 11 November 2013.
40. Newberry, K. (July 1988). "Production of a hybrid between the bedbugs Cimex hemipterus and Cimex lectularius". Medical and Veterinary Entomology. The Royal Entomological Society. 2 (3): 297–300. doi:10.1111/j.1365-2915.1988.tb00199.x. PMID 2980186.
41. Walpole, Debra E.; Newberry, K. (July 1988). "A field study of mating between two species of bedbug in northern KwaZulu, South Africa". Medical and Veterinary Entomology. The Royal Entomological Society. 2 (3): 293–296. doi:10.1111/j.1365-2915.1988.tb00198.x. PMID 2980185.
42. Oliver Otti; Klaus Reinhard; Aimee P McTighe (2013). "In Vitro Antimicrobial Sperm Protection by an Ejaculate-like Substance". Functional Ecology. 27 (1): 219–26. doi:10.1111/1365-2435.12025.
43. Klaus Reinhardt; Richard Naylor; Michael T Siva-Jothy (2011). "Male Mating Rate Is Constrained by Seminal Fluid Availability in Bedbugs, Cimex Lectularius". PLOS ONE. 6 (7): 1–8.
44. Mellanby, Kenneth. "Fertilization and Egg Production in the Bed-bug, Cimex Lectularius L." Parasitology 1939; 31(02): 193. Print.
45. Ryne Camille (2009). "Homosexual Interactions in Bed Bugs: Alarm Pheromones as Male Recognition Signals". Animal Behaviour. 78 (6): 1471–475. doi:10.1016/j.anbehav.2009.09.033.
46. Davies, Nicholas B.; Krebs, John R.; West, Stuart A. (17 February 2012). An Introduction to Behavioural Ecology (Kindle Locations 5432-5434). Wiley. Kindle Edition.
47. Xavier Bonnefoy; Helge Kampen; Kevin Sweeney. "Public Health Significance of Urban Pests" (PDF). World Health Organization. p. 136. Retrieved 1 December 2016.

48. Shukla; Upadhyaya (2009). Economic Zoology (Fourth ed.). Rastogi. p. 73. ISBN 978-81-7133-876-4.
49. Coughlan, Sean (29 January 2014). "Catching the mother of all bed bugs". BBC. Retrieved 29 January 2014.
50. Reinhardt, Klaus; Siva-Jothy, Micheal (2007). "Biology of the Bed Bugs (Cimicidae)". Annual Review of Entomology. 52: 351–374. doi:10.1146/annurev.ento.52.040306.133913. PMID 16968204. Retrieved 10 July 2014.
51. Reis Matthew D., Miller Dini M. (2011). "Host Searching and Aggregation Activity of Recently Fed and Unfed Bed Bugs (Cimex Lectularius L.)". Insects. 2 (2): 186–94.
52. Margie Pfiester; Philip G. Koehler; Roberto M. Pereira (2009). "Effect of Population Structure and Size on Aggregation Behavior Of(Hemiptera: Cimicidae)". Journal of Medical Entomology. 46 (5): 1015–020. doi:10.1603/033.046.0506.
53. Steelman, C.D. 2000. Biology and control of bed bugs, Cimex lectularius, in poultry houses. Avian Advice 2: 10,15.
54. Susan L. Woodward; Joyce A. Quinn (30 September 2011). Encyclopedia of Invasive Species: From Africanized Honey Bees to Zebra Mussels: From Africanized Honey Bees to Zebra Mussels. ABC-CLIO. p. 124. ISBN 978-0-313-38221-5. Retrieved 15 August 2013.
55. Anderson, AL; Leffler, K (May 2008). "Bedbug infestations in the news: a picture of an emerging public health problem in the United States" (PDF). Journal of environmental health. 70 (9): 24–7, 52–3. PMID 18517150.
56. Doggett, SL; Dwyer, DE; Peñas, PF; Russell, RC (January 2012). "Bed bugs: clinical relevance and control options.". Clinical Microbiology Reviews. 25 (1): 164–92. doi:10.1128/cmr.05015-11. PMC 3255965 . PMID 22232375.
57. "7 On Your Side: Get rid of bed bugs for less than $15". Retrieved 1 December 2016.
58. "Detecting Bed Bugs Using Bed Bug Monitors (from Rutgers NJAES)". Retrieved 1 December 2016.
59. Hulasare, Raj. "Fundamental Research on the Efficacy of Heat on Bed Bugs and Heat Transfer in Mattresses.".
60. 2012, Scientific American, Hayes
61. York Times. In Search of a Bedbug Solution. Published: 4 September 2010.
62. Miller, Dini (2008). "Bed bugs (hemiptera: cimicidae: Cimex spp.)". In Capinera, John L. Encyclopedia of Entomology (Second ed.). Springer. p. 414. ISBN 978-1-4020-6242-1.
63. "Natural Fungus May Provide Effective Bed Bug Control".
64. Roberts, Michelle. "Bed-bug DNA scanned for vulnerabilities". Retrieved 4 February 2016.
65. Hahn, Jeffrey (November 1999). "Masked hunters". Yard & Garden Briefs. University of Minnesota Extension Service: Yard & Garden Clinic. Archived

from the original on 23 March 2010. Retrieved 27 May 2010.

66. Encyclopedia Americana, 1996 ed., v. 3, p. 413
67. Heukelbach, J; Hengge, UR (2009). "Bed bugs, leeches and hookworm larvae in the skin". Clinics in dermatology. 27 (3): 285–90. doi:10.1016/j.clindermatol.2008.10.008. PMID 19362691
68. Romero A, Potter MF, Potter DA, Haynes KF (2007). "Insecticide Resistance in the Bed Bug: A Factor in the Pest's Sudden Resurgence?". Journal of medical entomology. 22 (2): 175–178. doi:10.1603/0022-2585(2007)44[175:IRITBB]2.0.CO;2 (inactive August 2016). ISSN 0022-2585. PMID 17427684.
69. Owen, James (13 May 2004). "Bloodthirsty Bedbugs Stage Comeback in U.S., Europe". National Geographic News. National Geographic. Retrieved 31 May 2010.
70. Bankhead, Charles (27 August 2015). "Bed Bug Resurgence a Multifactorial Issue: Hygiene, insecticide bans, globalization all contribute". Meeting Coverage. MedPage Today. Archived from the original on 28 August 2015. Retrieved 28 August 2015.
71. Krause-Parello CA, Sciscione P (April 2009). "Bedbugs: an equal opportunist and cosmopolitan creature". J Sch Nurs. 25 (2): 126–32. doi:10.1177/1059840509331438. PMID 19233933.
72. Xavier Bonnefoy; Helge Kampen; Kevin Sweeney. "Public Health Significance of Urban Pests" (PDF). World Health Organization. p. 131. Retrieved 1 December 2016.
73. Cranshaw, W.S.; Camper, M.; Peairs, F.B. (Feb 2009). "Bat Bugs and Bed Bugs". Colorado State University Extension. Retrieved 27 May 2010.
74. Weather.com: Spotted in Florida For First Time Since WWII, https://weather.com/news/trending/video/florida-fears-bedbug-infestation?pl=pl-editor-picks, Posted: 14 November 2016 04:45 PM EST, Updated: 14 November 2016 04:45 PM EST -- "Tropical bedbugs invade Florida for the first time in 60 years. There are now fears of a widespread infestation."
75. Campbell Brittany E.; Koehler Philip G.; Buss Lyle J.; Baldwin Rebecca W. (2016). "Recent Documentation of the Tropical Bed Bug (Hemiptera: Cimicidae) in Florida Since the Common Bed Bug Resurgence". Florida Entomologist. 99 (3): 549–551. doi:10.1653/024.099.0333.
76. A dictionary of Greek and Roman antiquities – Sir William Smith – Google Boeken. Books.google.com. Retrieved 11 November 2013.
77. John Southall. "That soon after the Fire of London, in some of the new-built Houses they were observ'd to appear, and were never noted to have been seen in the old, tho' they were then so few, as to be little taken notice of; yet as they were only seen in Firr-Timber, 'twas conjectured they were then first brought to England in them; of which most of the new Houses were partly built, instead of the good Oak destroy'd in the old.". A Treatise of Buggs [sic], pp. 16–17. Retrieved 1 December 2016.

78. Johann Friedrich Wolff; Johann Philip Wolff. "According to Scopoli's 2nd work (loc. cit.), found in Carniola and adjoining regions. According to Linnaeus' second work on exotic insects (loc. cit.), before the era of health, already in Europe, seldom observed in England before 1670.". Icones Cimicum descriptionibus illustratae. p. 127. Retrieved 1 December 2016. fourth fascicle (1804)

79. George Orwell, Down and Out in Paris and London, 1933

80. Schaefer, C.W.; Pazzini, A.R. (28 July 2000). Heteroptera of Economic Importance. Boca Raton, FL: CRC Press. p. 525. ISBN 0-8493-0695-7.

81. Kambu, Kabangu; Di Phanzu, N.; Coune, Claude; Wauters, Jean-Noël; Angenot, Luc (1982). "Contribution à l'étude des propriétés insecticides et chimiques d'Eucalyptus saligna du Zaïre (Contribution to the study of insecticide and chemical properties of Eucalyptus saligna from Zaire (Congo))". Plantes Médicinales et Phytothérapie. Paris: Jouve. 16 (1): 34–38. hdl:2268/14438.

82. "Getting Rid of Bed-Bugs". Grubstreet.rictornorton.co.uk. Retrieved 11 November 2013.

83. "Icones Cimicum descriptionibus illustratae". Archive.org. Retrieved 11 November 2013.

84. (no byline) (17 June 1848). "Peat and peat mosses". Scientific American. 3 (39): 307. Retrieved 26 May 2010.

85. Hill, Stuart B. (May 1986). "Diatomaceous Earth: A Non Toxic Pesticide". Macdonald J. Ste-Anne de Bellevue, QC: Macdonald College. 47 (2): 14–42. Archived from the original on 26 May 2010. Retrieved 26 May 2010.

86. Michael F. Potter; Kenneth F. Haynes; Chris Christensen; T. J. Neary; Chris Turner; Lawrence Washburn; Melody Washburn (December 2013). "Diatomaceous Earth: Where Do Bed Bugs Stand When the Dust Settles?". PCT Magazine (12): 72. ISSN 0730-7608.

87. Michael F. Potter; Kenneth F. Haynes; Jennifer R. Gordon; Larry Washburn; Melody Washburn; Travis Hardin (August 2014). "Silica Gel: A Better Bed Bug Desiccant". PCT Magazine (8): 76. ISSN 0730-7608.

88. Boase, C. (2001). "Bedbugs – back from the brink". Pesticide Outlook. 12 (4): 159–162. doi:10.1039/b106301b. Retrieved 27 May 2010.

89. Szyndler, M.W.; Haynes, K.F.; Potter, M.F.; Corn, R.M.; Loudon, C. (2013). "Entrapment of bed bugs by leaf trichomes inspires microfabrication of biomimetic surfaces" (PDF). Journal of The Royal Society Interface. 10 (83): 20130174. doi:10.1098/rsif.2013.0174. ISSN 1742-5662.

90. Boase, Clive J. (April 2004). "Bed-bugs – reclaiming our cities". Biologist. 51: 1–4. Retrieved 7 June 2010.

91. Potter, Michael F. (2011). "The History of Bed Bug Management – With Lessons from the Past" (PDF). American Entomologist.

92. Gerberg, Eugene J. (16 November 2008). "Entomologists in World War II" (PDF). Proceedings of the DOD Symposium, 'Evolution of Military Medical Entomology', Held 16 November 2008, Reno, NV. Annual Meeting of the

Entomological Society of America. Armed Forces Pest Management Board. Retrieved 11 November 2013.

93. Newsweek (8 September 2010). "The Politics of Bedbugs". Archived from the original on 21 October 2010. Retrieved 28 October 2010.
94. Voiland, Adam (16 July 2007). "You May not be Alone". U.S. News & World Report. 143 (2): 53–54. Archived from the original on 7 November 2011.
95. Megan Gibson (19 August 2010). "Are Bedbugs Taking Over New York City?". Time Magazine.
96. Metropolitan Tenants Organization (16 July 2013). "Chicago Council passes Bed Bug Ordinance". Metropolitan Tenants Organization website.
97. Faúndez E. I., Carvajal M. A. "Bed bugs are back and also arriving is the southernmost record of Cimex lectularius (Heteroptera: Cimicidae) in South America". Journal of Medical Entomology. 51 (5): 1073–1076. doi:10.1603/me13206.
98. Faúndez E. I. (2015). "Primeros registros de la chinche de cama Cimex lectularius Linneo, 1755 (Hemiptera: Cimicidae) en la Isla Tierra del Fuego (Chile)". Arquivos Entomolóxicos. 14: 279–280.
99. Austin, James (2008). "Bed Bugs". Urban and Structural Pests. Center for Urban & Structural Entomology, Department of Entomology, Texas A&M. Archived from the original on 25 June 2010. Retrieved 31 May 2010.
100. Steelman, C.D.; Szalanski, A.L.; Trout, R.; McKern, J.A.; Solorzano, C.; Austin, J.W. (2008). "Susceptibility of the bed bug Cimex lectularius L. (Hemiptera: Cimicidae) to selected insecticides". Journal of Agricultural and Urban Entomology. 25 (1): 41–51. doi:10.3954/1523-5475-25.1.41.
101. Berg, Rebecca (2010). "Bed Bugs: The Pesticide Dilemma". Journal of Environmental Health. 72 (10): 32–35. PMID 20556941.
102. Stephen L Doggett; Dominic E. Dwyer; Richard C Russell (January 2012). "Bed Bugs ClinicalRelevance and Control Options". Clinical Microbiology Review. 25: 164–92. doi:10.1128/CMR.05015-11. PMC 3255965 . PMID 22232375.
103. "Online Etymology Dictionary". Etymonline.com. Retrieved 11 November 2013.
104. The Oxford English Dictionary, 2nd edition., quotes sources using all three spellings
105. de la Fuente (2003)
106. Dunlop, Jason A.; et al. (2016). "Microtomography of the Baltic amber tick Ixodes succineus reveals affinities with the modern Asian disease vector Ixodes ovatus". BMC Evolutionary Biology. 16. doi:10.1186/s12862-016-0777-y.
107. Guglielmone et al. (2010)
108. Goddard (2008): p. 80
109. "Systematics and evolution of ticks with a list of valid genus and species names". Parasitology. 129: S15–S36. 2004. doi:10.1017/S0031182004005207.
110. Klompen, J.S.; Black, W.C.; Keirans, J.E.; Oliver, J.H. (1996). "Evolution

of ticks". Annual Review of Entomology. 41: 141–61. doi:10.1146/annurev.ento.41.1.141.

111. Anderson John F (2002). "The natural history of ticks". Medical Clinics of North America. 86 (2): 205–218. doi:10.1016/s0025-7125(03)00083-x.
112. Crampton, A.; McKay, I.; Barker, S. C. (May 1996). "Phylogeny of ticks (Ixodida) inferred from nuclear ribosomal DNA". Int. J. Parasitology. 26 (5): 511–517. doi:10.1016/0020-7519(96)89379-9. PMID 8818731.
113. Black, William C., IV; Piesman, Joseph (October 1994). "Phylogeny of hard- and soft-tick taxa (Acari: Ixodida) based on mitochondrial 16S rDNA sequences" (PDF). PNAS. 91: 10034–10038. doi:10.1073/pnas.91.21.10034.
114. Molyneux (1993) p. 6
115. Keirans et al. (1976)
116. Gu, X. B.; et al. (2014). "The complete mitochondrial genome of the scab mite Psoroptes cuniculi (Arthropoda: Arachnida) provides insights into Acari phylogeny". Parasit Vectors. 7 (340). doi:10.1186/1756-3305-7-340. PMC 4223567 . PMID 25052180.
117. Magnarelli (2009)
118. Nuttall (1905)
119. Dantas-Torres, Filipe; Edmilson F. Oliveira-Filho; Fábio Ângelo M. Soares; Bruno O.F. Souza; Raul Baltazar P. Valença; Fabrício B. Sá (2008). "Ticks Infesting Amphibians and Reptiles in Pernambuco, Northeastern Brazil" (PDF). Rev. Bras. Parasitol. Vet. 17 (4): 218–221.
120. "Ticks of Livestock". Ectoparasites of Livestock. Butox. Retrieved 14 January 2017.
121. "Ticks". Livestock Veterinary Entomology. Texas A&M AgriLife. Retrieved 14 January 2017.
122. Hoogstraal, Harry; Kaiser, Makram N.; Traylor, Melvin A.; Guindy, Ezzat; Gaber, Sobhy (1963). "Ticks (Ixodidae) on birds migrating from Europe and Asia to Africa, 1959-61". Bulletin of the World Health Organization. 28 (2): 235–262. PMC 2554471 . PMID 13961632.
123. b Wall & Shearer (2001): p. 55
124. Allan (2001)
125. Ruppert, Edward E.; Fox, Richard, S.; Barnes, Robert D. (2004). Invertebrate Zoology, 7th edition. Cengage Learning. pp. 590–595. ISBN 978-81-315-0104-7.
126. "Common Ticks". Illinois Department of Public Health. Retrieved 11 April 2014.
127. "Soft ticks". CVBD: Companion Vector-Borne Diseases. Retrieved 6 December 2016.
128. Sonenshine (2005): p. 14
129. Nicholson et al. (2009): p. 486
130. For Haller's organ, see also: Mehlhorn (2008): p. 582.

131. Walker, Jane B.; Keirans, James E.; Horak, Ivan G. (2005). The Genus Rhipicephalus (Acari, Ixodidae): A Guide to the Brown Ticks of the World. Cambridge University Press. p. 39. ISBN 978-1-316-58374-6.

132. Salman, Mowafak Dauod; Tarrés-Call, Jordi; Estrada-Peña, Agustín (2013). Ticks and Tick-borne Diseases: Geographical Distribution and Control Strategies in the Euro-Asia Region. CABI. pp. 6–12. ISBN 978-1-84593-853-6.

133. "Soft ticks". CVBD: Companion Vector-Borne Diseases. Retrieved 6 December 2016.

134. Roshdy et al. (1983)

135. Brouwers, Lucas (30 August 2011). "Long Lost Relative of Ticks Pops Up Again". Scientific American. Retrieved 4 December 2016.

136. "Life cycle of Hard Ticks that Spread Disease". Centers for Disease Control and Prevention. Retrieved 22 June 2013.

137. Klompen and Grimaldi (2001):

138. #KlompenGrimaldi

139. Goddard (2008): p. 82

140. Mans, Luow, and Neitz (2002):

141. "Host seeking". CVBD: Companion Vector-Borne Diseases. Retrieved 8 December 2016.

142. "Hard ticks". CVBD: Companion Vector-Borne Diseases. Retrieved 6 December 2016.

143. Ray, C. Claiborne (28 May 2012). "The mighty tick". New York Times. Retrieved 15 December 2016.

144. "Disease transmission". CVBD: Companion Vector-Borne Diseases. Retrieved 9 December 2016.

145. Dennis & Piesman (2005): p. 5

146. Aeschlimann & Freyvogel, 1995: p. 182

147. Unsworth, N.B.; Stenos, J; Graves, S.R. et al. (2007). "Flinders Island spotted fever rickettsioses caused by "marmionii" strain of Rickettsia honei, Eastern Australia". Emerging Infectious Diseases. 13 (4): 566–73. doi:10.3201/eid1304.050087. PMC 2725950 . PMID 17553271.

148. "Q fever". Centers for Disease Control. Retrieved November 7, 2010.

149. "Heartland virus".

150. Feldman, Getty; Feldman, Lauree; Macnair, Patricia (reviewer) (6 March 2014). "All about ticks: why it's so important to remove a tick". Netdoctor. Retrieved 8 December 2016.

151. "Correct tick removal". Borreliosis and Associated Diseases Awareness UK. Retrieved October 24, 2014.

152. "Tick removal". Lyme Disease Action. Retrieved October 24, 2014.

153. "Tick removal". Centers for Disease Control and Prevention. Retrieved October

24, 2014.

154. "Catalyst: Tick allergy - ABC TV Science". www.abc.net.au. 17 February 2015. Retrieved 2015-12-10.

155. "Tick Removal". Centers for Disease Control and Prevention. 1 June 2015. Retrieved 15 January 2017.

156. Dennis & Piesman, 2005: p. 3

157. Plantard, O.; et al. (2012). "Detection of Wolbachia in the tick Ixodes ricinus is due to the presence of the hymenoptera endoparasitoid Ixodiphagus hookeri". PLOS ONE. 7 (1): e30692. doi:10.1371/journal.pone.0030692.

158. Tijsse-Klasen, Ellen; et al. (2011). "Parasites of vectors - Ixodiphagus hookeri and its Wolbachia symbionts in ticks in the Netherlands". Parasites & Vectors. 4 (228). doi:10.1186/1756-3305-4-228.

159. Duffy et al. (1992)

160. "Hartz flea and tick drops for cats and kittens to be phased out". Environmental Protection Agency. Archived from the original on January 11, 2010. Retrieved September 16, 2011

CHAPTER 6

Economic Entomology

Chapter Contents

Economic Entomology, the name given to the study of insects based on their relation to man, his domestic animals and his crops, and, in the case of those that are injurious, of the practical methods by which they can be prevented from doing harm, or be destroyed when present. In Great Britain little attention is paid to this important branch of agricultural science, but in America and the British colonies the case is different. Nearly every state in America has its official economic entomologists, and nearly every one of the British crown colonies is provided with one or more able men who help the agricultural community to battle against the insect pests. Most, if not all, of the important knowledge of remedies comes from America, where this subject reaches the highest perfection; even the life-histories of some of the British pests have been traced out in the United States and British colonies more completely than at home, from the creatures that have been introduced from Europe.

INTRODUCTION

Economic entomology is a field of entomology, which involves the study of insects that benefit or harm humans, domestic animals, and crops. Insects that cause losses are termed as pests. Some species can cause indirect damage by spreading diseases and these are termed as vectors. Those that are beneficial include those reared for food such as honey, substances such as lac or pigments and for their role in pollinating crops and controlling pests.

HISTORY

In the 18th century many works were published on agriculture. Many contained accounts of pest insects. In France Claude Sionnest (1749–1820) was a notable figure.

19th Century

The most able exponent of this subject in Great Britain was John Curtis, whose treatise *Farm Insects*, published in 1860, was once the standard British work dealing with the insect pests of corn, roots, grass and stored corn. The most important works dealing with fruit and other pests were by Saunders, Joseph Albert Lintner, Charles Valentine Riley, Mark Vernon Slingerland and others in America and Canada. In Europe the earliest works were by Ernst Ludwig Taschenberg, Sven Lampa (1839–1914), Enzio Reuter (1867–1951) and Vincenze Kollar. Charles French (1842–1933), Walter Wilson Froggatt (1858–1937) and Henry Tryon (1856–1943) pioneered in Australia. It was not until the last quarter of the 19th century that any real advance was made in the study of economic entomology. Among the early writings, besides the book of Curtis, there was also a publication by Pohl and Kollar, entitled *Insects Injurious to Gardeners, Foresters and Farmers*, published in 1837, and Taschenberg's *Praktische Insecktenkunde*. During the 19th century Italian entomologists made significant progress in controlling diseases of the Silk moth which supported the silk industry, in the control of agricultural pests and in stored product entomology. Significant figures were: Agostino Bassi (1773–1856), Camillo Róndani (1808–1879), Adolfo Targioni Tozzetti (1823–1902), Pietro Stefanelli (1835, 1919), Camillo Acqua (1863–1936) Antonio Berlese (1863–1927), Gustavo Leonardi(1869–1918) and Enrico Verson (1845–1927). In France Etienne Laurent Joseph Hippolyte Boyer de Fonscolombe, Charles Jean-Baptiste Amyot, Émile Blanchard, Valéry Mayet and Claude Charles Goureau were early workers, as was Jean Victoire Audouin, the author of *Histoire des insectes nuisibles à la vigne et particulièrement de la Pyrale*, Philippe Alexandre Jules Künckel d'Herculais and Jean-Étienne Girard. American literature began as far back as 1788, when a report on the Hessian fly was issued by Sir Joseph Banks; in 1817 Thomas Say began his writings; while in 1856 Asa Fitch started his report on *Noxious Insects of New York*. Also in America, Matthew Cooke wrote *Treatise on the Insects Injurious to Fruit and Fruit Trees of the State of California, and Remedies Recommended for Their Extermination*, published in 1881. The Englishman Frederick Vincent Theobald wrote A textbook of agricultural zoology in 1890. It became a standard text worldwide. Notable foresters were Herman von Nördlinger (1818–1897) and Julius Theodor Christian Ratzeburg (1801–1871)

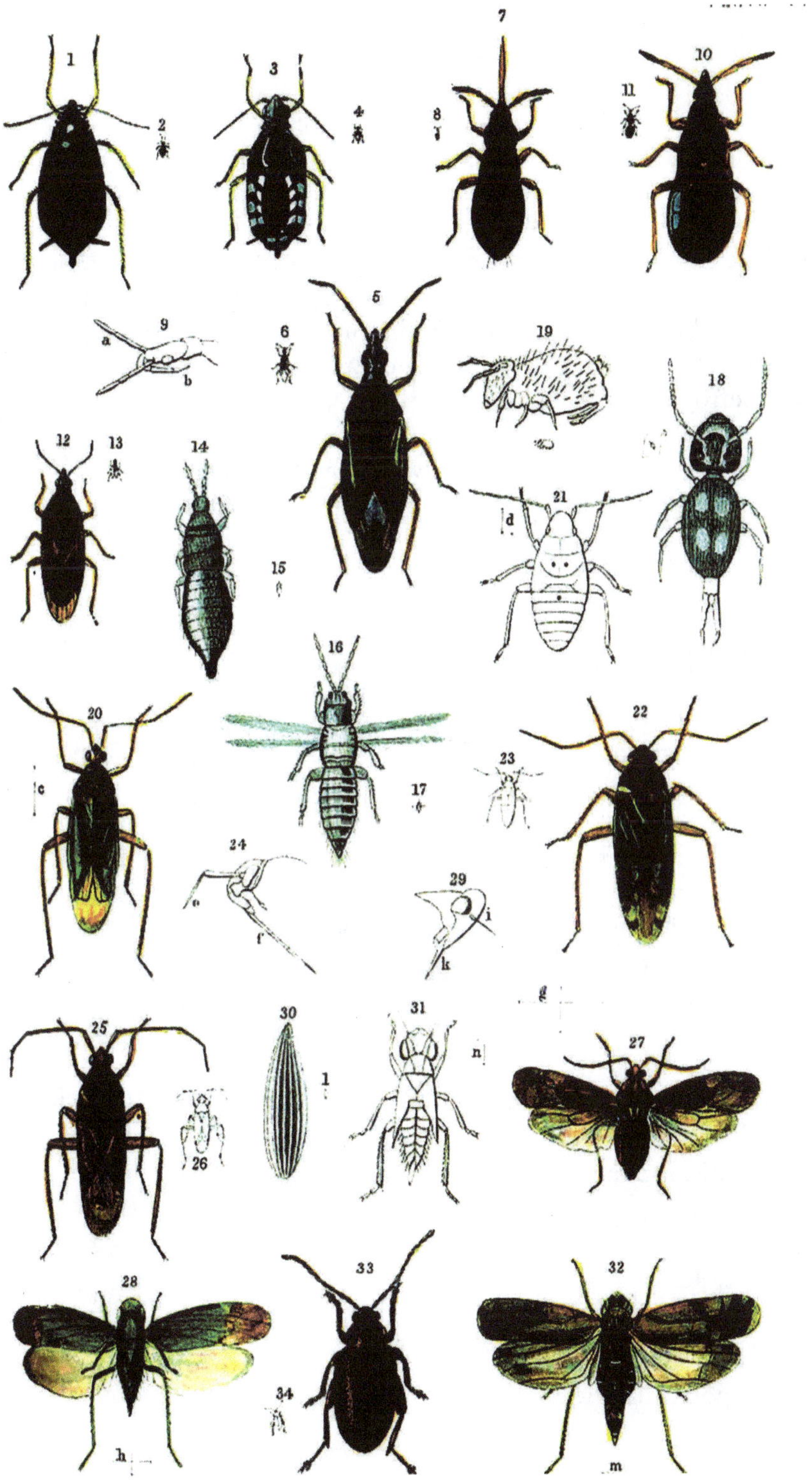

"Insects infesting potato crops": a plate from John Curtis's *Farm Insects*, 1860.

20th Century

Among the most important reports early in the 20th century were those of Charles Valentine Riley, published by the U.S. Department of Agriculture, extending from 1878 to his death, in which is embodied an enormous amount of valuable material. At his death the work fell to Professor Leland Ossian Howard, in the form of *Bulletin of the U.S. Department of Agriculture*. The chief writings of J. A. Lintner extend from 1882 to 1898, in yearly parts, under the title of *Reports on the Injurious Insects of the State of New York*. Another significant contributor to the entomological literature of the United States was Charles W. Woodworth. The Florida entomologist Wilmon Newell was a pioneer of pest control as was Clarence Preston Gillette. In India Thomas Bainbrigge Fletcher, who succeeded Harold Maxwell-Lefroy and Lionel de Nicéville as the first Imperial Entomologist, wrote *Some South Indian insects and other animals of importance considered especially from an economic point of view*, an influential work in the subcontinent. In France Alfred Balachowsky was a key figure. In the last quarter of the 20th century new techniques were pioneered and new theories developed, for instance Integrated Pest Management by Ray F. Smith.

HARMFUL INSECTS

Insects considered *pests* of some sort occur among all major living orders with the exception of Ephemeroptera (mayflies), Odonata, Plecoptera (stoneflies), Embioptera (webspinners), Trichoptera (caddisflies), Neuroptera (in the broad sense), and Mecoptera (also, the tiny groups Zoraptera, Grylloblattodea, and Mantophasmatodea). Conversely, of course, essentially all insect orders primarily have members which are beneficial, in some respects, with the exception of Phthiraptera (lice), Siphonaptera (fleas), and Strepsiptera, the three orders whose members are exclusively parasitic.

Insects are considered as pests for a variety of reasons including their

- direct damage by feeding on crop plants in the field or by infesting stored products
- indirect damage by spreading viral diseases of crop plants (especially by sucking insects such as leafhoppers)
- spreading disease among humans and livestock
- annoyance to humans

Examples

- The Phylloxera plague
- Migratory locust
- Colorado potato beetle
- Boll weevil
- Japanese beetle
- Aphids
- Mosquitoes

The phylloxera, a true gourmet, finds out the best vineyards and attaches itself to the best wines Cartoon from Punch, 6 September 1890)

- Cockroach
- Western corn rootworm
- Some fly species

In the past entomologists working on pest insects attempted to *eradicate* species. This has rarely worked except in islands or controlled environments and raises ethical issues. Over time the language has changed to terms like *control* and *management*. The indiscriminate use of toxic and persistent chemicals and the resurgence of pests in the history of cotton growing in the US has been particularly well studied.

BENEFICIAL INSECTS

Honey is perhaps the most economically valuable product from insects. Apiculture is a commercial enterprise in most parts of the world and many forest tribes have been dependent on honey as a major source of nutrition. Honeybees can also act as pollinators of crop species. Many predators and parasitoid insects are encouraged and augmented in modern agriculture.

Silk is extracted from both reared caterpillars as well as from the wild (producing wild silk). Sericulture deals with the techniques for efficient silkworm rearing and silk production. Although new fabric materials have substituted silk in many applications, it continues to be the material of choice for surgical sutures.

Boll Weevil Monument, erected by the citizens of Enterprise, Alabama to honour the pest that ended their dependence on cotton, a poverty crop.

Lac was once extracted from scale insects but is now replaced by synthetic substitutes. The dye extracted from cochineal insects was similarly replaced by technological advances.

The idea of insects as human food, entomophagy, has been proposed as a solution to meet the growing demand for food, but has not gained widespread acceptance.

Phylloxera

Grape phylloxera (*Daktulosphaira vitifoliae* (Fitch 1855); family Phylloxeridae); originally described in France as *Phylloxera vastatrix*; equated to the previously described *Daktulosphaira vitifoliae*, *Phylloxera vitifoliae*; commonly just called **phylloxera** is a pest of commercial grapevines worldwide, originally native to eastern North America.

These almost microscopic, pale yellow sap-sucking insects, related to aphids, feed on the roots and leaves of grapevines (depending on the phylloxera genetic strain). On *Vitis vinifera L.*, the resulting deformations on roots ("nodosities" and "tuberosities") and secondary fungal infections can girdle roots, gradually cutting off the flow of nutrients and water to the vine. Nymphs also form protective galls on the undersides of grapevine leaves of some *Vitis* species and overwinter under the bark or on the vine roots; these leaf galls are typically only found on the leaves of American vines.

American vine species (such as *Vitis labrusca*) have evolved to have several natural defenses against phylloxera. The roots of the American vines exude a sticky sap that repels the nymph form when it tries to feed from the vine by clogging its mouth. If the nymph is successful in creating a feeding wound on the root, American vines respond by forming a protective layer of tissue to cover the wound and protect it from secondary bacterial or fungal infections.

Currently there is no cure for phylloxera and unlike other grape diseases such as powdery or downy mildew, there is no chemical control or response. The only successful means of controlling phylloxera has been the grafting of phylloxera resistant American rootstock (usually hybrid varieties created from the *Vitis berlandieri*, *Vitis riparia* and *Vitis rupestris* species) to more susceptible European *vinifera* vines.

Biology

Phylloxera nymphs feeding on the roots.

The phylloxera aphid has a complex life-cycle of up to 18 stages, that can be divided into four principal forms: sexual form, leaf form, root form, and winged form.

The **sexual form** begins with male and female eggs laid on the underside of young grape leaves. The male and female at this stage lack a digestive system, and once hatched, they mate and then die. Before the female dies, she lays one winter egg in the bark of the vine's trunk. This egg develops into the **leaf form**. This nymph, the fundatrix (stem mother), climbs onto a leaf and lays eggs parthenogenetically in a leaf gall that she creates by injecting saliva into the leaf. The nymphs that hatch from these eggs may move to other leaves, or move to the roots where they begin new infections in the **root form**. In this form they perforate the root to find nourishment, infecting the root with a poisonous secretion that stops it from healing. This poison eventually kills the vine. This nymph reproduces by laying eggs for up to seven more generations (which also can reproduce parthenogenetically) each summer. These offspring spread to other roots of the vine, or to the roots of other vines through cracks in the soil. The generation of nymphs that hatch in the autumn hibernate in the roots and emerge next spring when the sap begins to rise. In humid areas, the nymphs develop into the **winged form**, else they perform the same role without wings. These nymphs start the cycle again by either staying on the vine to lay male and female eggs on the bottom side of young grape leaves, or flying to an uninfected vine to do the same.

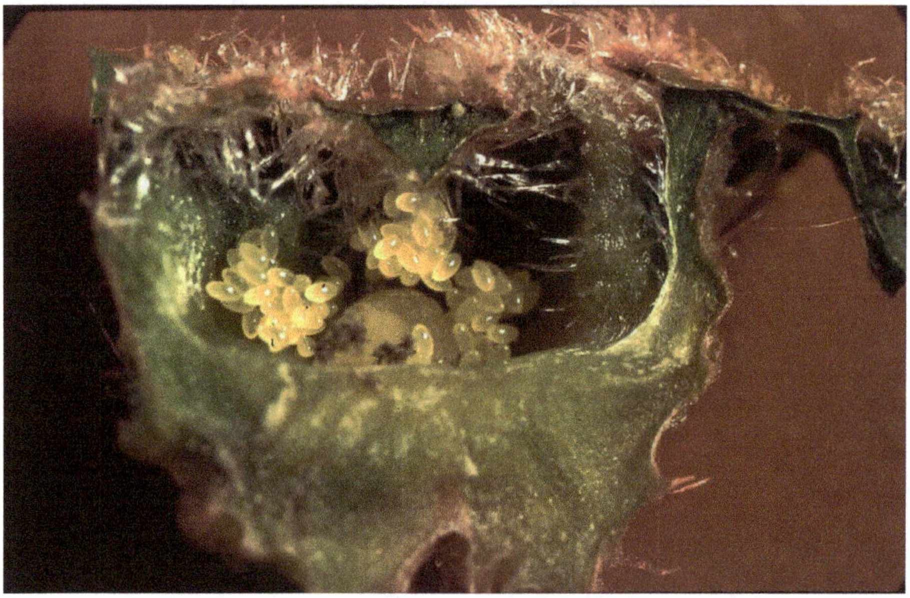

Phylloxera eggs inside a leaf gall.

Many attempts have been made to interrupt this life cycle to eradicate phylloxera, but the louse has proven to be extremely adaptable, as no one stage of the life cycle is solely dependent upon another for the propagation of the species.

Fighting the "Phylloxera Plague"

In the late 19th century the phylloxera epidemic destroyed most of the vineyards for wine grapes in Europe, most notably in France. Phylloxera was introduced to Europe when avid botanists in Victorian England collected specimens of American vines in the 1850s. Because phylloxera is native to North America, the native grape species are at least partially resistant. By contrast, the European wine grape *Vitis vinifera* is very susceptible to the insect. The epidemic devastated vineyards in Britain and then moved to the European mainland, destroying most of the European grape growing industry. In 1863, the first vines began to deteriorate inexplicably in the southern Rhône region of France. The problem spread rapidly across the continent. In France alone, total wine production fell from 84.5 million hectolitres in 1875 to only 23.4 million hectolitres in 1889. Some estimates hold that between two-thirds and nine-tenths of all European vineyards were destroyed.

In France, one of the desperate measures of grape growers was to bury a live toad under each vine to draw out the "poison". Areas with soils composed principally of sand or schist were spared, and the spread was slowed in dry climates, but gradually the aphid spread across the continent. A significant amount of research was devoted to finding a solution to the phylloxera problem, and two major solutions gradually emerged: grafting cuttings onto resistant rootstocks and hybridization.

Response

François Baco, creator of Baco blanc, was one of many grape breeders to introduce hybrid wine grape varieties in response to the phylloxera epidemic.

By the end of the 19th Century, **hybridization** became a popular avenue of research for stopping the phylloxera louse. Hybridization is the breeding of *Vitis vinifera* with resistant species. Most native American grapes are naturally phylloxera resistant (*Vitis aestivalis*, *rupestris*, and *riparia* are particularly so, while *Vitis labrusca* has a somewhat weak resistance to it) but have aromas that are off-putting to palates accustomed to European grapes. The intent of the cross was to generate a hybrid vine that was resistant to phylloxera but produced wine that did not taste like the American grape. Ironically, the hybrids tend not to be especially resistant to phylloxera, although they are much more hardy with respect to climate and other vine diseases. The new hybrid varieties have never gained the popularity of the traditional ones. In the EU they are generally banned or at least strongly discouraged from use in quality wine, although they are still in widespread use in much of North America, such as Missouri, Ontario, and upstate New York.

Grafting with Resistant Rootstock

A grape leaf showing the galls that are formed during a phylloxera infestation.

Use of a resistant, or tolerant, rootstock, developed by Charles Valentine Riley in collaboration with J. E. Planchon and promoted by T. V. Munson, involved grafting a *Vitis vinifera* scion onto the roots of a resistant *Vitis aestivalis* or other American native species. This is the preferred method today, because the rootstock does not interfere with the development of the wine grapes (more technically, the genes responsible for the grapes are not in the rootstock but in the scion), and it furthermore allows the customization of the rootstock to soil and weather conditions, as well as desired vigor.

Not all rootstocks are equally resistant. Between the 1960s and the 1980s in California, many growers used a rootstock called AxR1. Even though it had already failed in many parts of the world by the early twentieth century, it was thought to be resistant by growers in California. Although phylloxera initially did not feed heavily on AxR1 roots, within twenty years, mutation and selective pressures within the phylloxera population began to overcome this rootstock, resulting in the eventual failure of most vineyards planted on AxR1. The replanting of afflicted vineyards continues today.

Many have suggested that this failure was predictable, as one parent of AxR1 is in fact a susceptible *V. vinifera* cultivar. But the transmission of phylloxera tolerance is more complex, as is demonstrated by the continued success of 41B, an F1 hybrid of *Vitis berlandieri* and *Vitis vinifera*. The full story of the planting of AxR1 in California, its recommendation, the warnings, financial consequences, and subsequent recriminations remains to be told. Modern phylloxera infestation also occurs when wineries are in need of fruit immediately, and choose to plant ungrafted vines rather than wait for grafted vines to be available.

Roots that have been damaged by phylloxera.

The use of resistant American rootstock to guard against phylloxera also brought about a debate that remains unsettled to this day: whether self-rooted vines produce better wine than those that are grafted. Of course, the argument is essentially irrelevant wherever phylloxera exists. Had American rootstock not been available and used, there would be no *V. vinifera* wine industry in Europe or most places other than Chile, Washington State, and most of Australia. Cyprus was spared by the phylloxera plague, and thus its wine stock has not been grafted for phylloxera resistant purposes.

Aftermath

A grafted vine with the scion (grape variety) visible as the darker wood above the graft union and the rootstock variety below.

The only European grapes that are natively resistant to phylloxera are the Assyrtiko grape which grows on the volcanic island of Santorini, Greece, although it is not clear whether the resistance is due to the rootstock itself or the volcanic ash on which it grows; and the Juan Garcia grape variety, autochthonous to the medieval village of Fermoselle in Spain. The Juan Garcia variety remained—untouched by phylloxera—sheltered on the vineyards planted on the man-made land terraces along the mountainous skirts on the gigantic and steep Arribes River Canyon, where the microclimatic conditions discourage the growth of phylloxera.

To escape the threat of phylloxera, wines have been produced since 1979 on the sandy beaches of Provence's Bouches-du-Rhône, which extends from the coastline of the Gard region to the waterfront village of Saintes Maries de la Mer. The sand, sun and wind in this area has been a major deterrent to phylloxera. The wine produced here is calledVins des Sables orwines of the sands. In the same department, where the canal irrigation system built by the Romans still partly persists to this day, winter flooding is also practiced where possible, for instance south of the city of Tarascon. Flooding the vineyards for 50 days kills all the nymphs that overwinter in the roots or the bark at the bottom of the plant.

Some regions were so blighted by phylloxera that they never recovered, and instead the producers switched crops entirely. The island of Mallorca is one example, where almonds now substitute for vines.

Vines that survived Phylloxera

A collection of vines with grafted rootstocks.

According to wine critic and author Kerin O'Keefe, thanks to tiny parcels of vineyards throughout Europe which were inexplicably unscathed, some vineyards still exist as they were before the phylloxera devastation.

So far, Chilean wine has remained phylloxera free. It is isolated from the world by the Atacama Desert to the north, the Pacific Ocean to the west, and the Andes Mountains to the east. Phylloxera has also never been found in several wine-growing regions of Australia, including Western Australia and South Australia. The Riesling of the Mosel region has also remained untouched by phylloxera; the parasite is unable to survive in the slate soil.

Three tiny parcels of ungrafted Pinot noir escaped phylloxera, which is used to produce Bollinger Vieilles Vignes Françaises, one of the rarest and most expensive Champagnes available.

A rare vintage port is made from ungrafted vines grown on a small parcel, called Nacional, in the heart of the Quinta do Noval estate. There is no scientific explanation as to why this plot survived while others succumbed.

Another vineyard unaffected by the phylloxera is the Lisini estate in Montalcino in Italy: a half-hectare vineyard of Sangiovese, with vines dating back to the mid-1800s. Since 1985 the winery has produced a few bottles of Prefillossero (Italian forbefore the phylloxera). The wine has devout followers, including Italian wine critic Luigi Veronelli who inscribed on a bottle of the 1987 at the winery that drinking Prefillossero was like listening to 'the earth singing to the sky'.

Jumilla in southeastern Spain is an important area of ungrafted vineyards, mainly from Monastrell grapes. Those vineyards, however, are not immune to the louse, which is slowly advancing and destroying the fabled Pie Franco vineyard of the Casa Castillo estate, planted in 1942, i.e., when phylloxera had already been in the region for five decades.

Large swaths of vineyards on the slopes of Sicily's volcano Mount Etna also remain free of the phylloxera louse. Some vines are more than one-hundred-fifty years old, pre-dating the phylloxera infestation in Sicily (1879–1880). Part of the reason for this is the high concentration of silica sand and very low (less than 3%) concentration of clay in the volcanic soils. In this environment (> 400 m AMSL) the surface water from heavy bouts of rain seals the soil so perfectly that it imprisons the louse, effectively drowning them before they are able to thrive.

REFERENCES

1. Shimer. 1866. The Prairie Farmer 18:36
2. Wine & Spirits Education TrustWine and Spirits: Understanding Wine Quality pgs 2-5, Second Revised Edition (2012), London, ISBN 9781905819157
3. McLeod, Murdick J.; Williams, Roger N.Grape Phylloxera. Archived from the original on 2012-07-16.
4. winepros.com.au The Oxford Companion to Wine.phylloxera.
5. Wines of the Sand. Feature Article. Novus Vinum. 2006-09-17. Retrieved 2008-11-05.
6. G. Gale.Saving the vine from phylloxera: a never-ending battle (PDF). University of Missouri-Kansas City.
7. O'Keefe, Kerin (October 2005).The great escape. Decanter.
8. Phylloxera. Vinehealth Australia. Retrieved 3 May 2016.
9. Pigott, Stuart.The Mosel River Renaissance. Los Angeles Times. Retrieved 3 May 2016.
10. Salcito, Jordan.Germany's Wine Revolution Is Just Getting Started. The Daily Beast. Campbell, Christy. "Phylloxera: How Wine Was Saved For the World. Harper Perennial, 2004, p. 129-130

CHAPTER 7

Myriapodology

Chapter Contents

- **Introduction**
- **Myriapoda**
- **Ecology**
- **Classification**

Definition of Myriapods in the Medical Dictionary. Myriapods explanation. Information about Myriapods in Free online English dictionary. What is Myriapods? Meaning of Myriapods medical term.

INTRODUCTION

Myriapodology is the scientific study of myriapods which includes centipedes and millipedes. The field of myriapodology can also cover other myriapods such as pauropods and symphylans. Those who study myriapods are myriapodologists.

MYRIAPODA

Myriapoda is a subphylum of arthropods containing millipedes, centipedes, and others. The group contains over 13,000 species, most of which are terrestrial.Although their name suggests they have myriad (10,000) legs, myriapods range from having over 750 legs (the millipede *Illacme plenipes*) to having fewer than ten legs.

The fossil record of myriapods reaches back into the late Silurian, although molecular evidence suggests a diversification in the Cambrian Period, and Cambrian fossils exist which resemble myriapods. The oldest unequivocal myriapod fossil is of the millipede *Pneumodesmus newmani*, from the late Silurian (428 million years ago). *P. newmani* is also important as the earliest known terrestrial animal. The phylogenetic classification of myriapods is still debated.

The scientific study of myriapods is myriapodology, and those who study myriapods are myriapodologists.

Anatomy

The head of *Scutigera coleoptrata*, showing antennae, compound eyes and mouthparts.

Myriapods have a single pair of antennae and, in most cases, simple eyes. The mouthparts lie on the underside of the head, with an "epistome and labrum forming the upper lip, and a pair of maxillae forming the lower lip. A pair of mandibles lie inside the mouth. Myriapods breathe through spiracles that connect to a tracheal system similar to that of insects. There is a long tubular heart that extends through much of the body, but usually few, if any, blood vessels.

Malpighian tubules excrete nitrogenous waste into the digestive system, which typically consists of a simple tube. Although the ventral nerve cord has a ganglion in each segment, the brain is relatively poorly developed.

During mating, male myriapods produce a packet of sperm, or spermatophore, which they must transfer to the female externally; this process is often complex and highly developed. The female lays eggs which hatch as much shortened versions of the adults, with only a few segments and as few as three pairs of legs. The young add additional segments and limbs as they repeatedly moult to reach the adult form.

ECOLOGY

Myriapods are most abundant in moist forests, where they fulfill an important role in breaking down decaying plant material,although a few live in grasslands, semi-arid habitats or even deserts. A very small percentage of species are littoral (found along the sea shore). The majority are detritivorous, with the exception of centipedes, which are chiefly nocturnal predators. Pauropodans and symphylans are small, sometimes microscopic animals that resemble centipedes superficially and live in soils. Millipedes differ from the other groups in having their body segments fused into pairs, giving the appearance that each segment bears two pairs of legs, while the other three groups have a single pair of legs on each body segment.

Although not generally considered dangerous to humans, many millipedes produce noxious secretions (often containing benzoquinones) which in rare cases can cause temporary blistering and discolouration of the skin. Large centipedes, however can bite humans, and although the bite may cause intense pain and discomfort, fatalities are extremely rare.

CLASSIFICATION

There has been much debate as to which arthropod group is most closely related to the Myriapoda. Under the Mandibulata hypothesis, Myriapoda is the sister taxon to Pancrustacea, a group comprising the Crustacea and Hexapoda (insects and their close relatives). Under the Atelocerata hypothesis, Hexapoda is the closest, whereas under the Paradoxopoda hypothesis, Chelicerata is the closest. This last hypothesis, although supported by few, if any, morphological characters, is supported by a number of molecular studies. There are four classes of extant myriapods, Chilopoda (centipedes), Diplopoda, Pauropoda and Symphyla, containing a total of around 12,000 species. While each of these groups of myriapods is believed to be monophyletic, relationships among them are less certain.

Centipedes

Scolopendra cingulata, a centipede.

Centipedes make up the class Chilopoda. They are fast, predatory and venomous, hunting mostly at night. There are around 3,300 species, ranging from the diminutive *Nannarrup hoffmani* (less than 12 mm or $\frac{1}{2}$ in in length) to the giant *Scolopendra gigantea*, which may exceed 30 centimetres (12 in).

Millipedes

Tachypodoiulus niger, a millipede.

Millipedes form the class Diplopoda. Most millipedes are slower than centipedes, and feed on leaf litter and detritus. They are distinguished by the fusion of each pair of body segments into a single unit, giving the appearance of having two pairs of legs per segment. Around 8,000 species have been described, which may represent less than a tenth of the true global millipede diversity. The name "millipede" is a compound word formed from the Latin roots *millia* ("thousand") and *pes* (gen. *pedis*) ("foot"), although millipedes typically have between 36 and 400 legs. One species, *Illacme plenipes*, has the greatest number of legs of any animal, with 750. Pill millipedes are much shorter, and are capable of rolling up into a ball, like pillbugs.

Symphyla

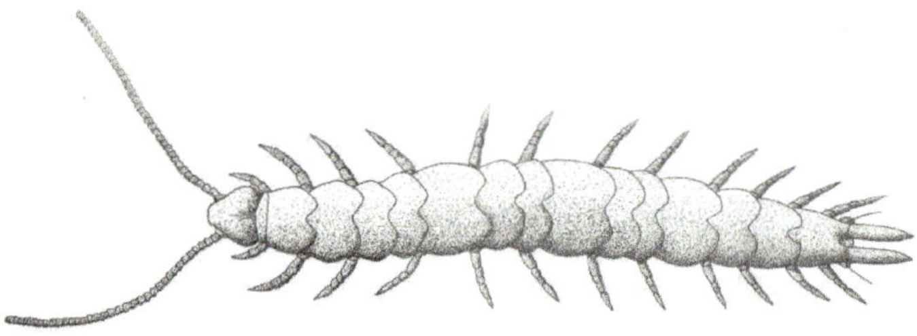

Scutigerella immaculata, a symphylan.

About 200 species of them are known worldwide. They resemble centipedes but are smaller and translucent. Many spend their lives as soil infauna, but some live arboreally. Juveniles have six pairs of legs, but, over a lifetime of several years, add an additional pair at each moult so that the adult instar has twelve pairs of legs.

Pauropoda

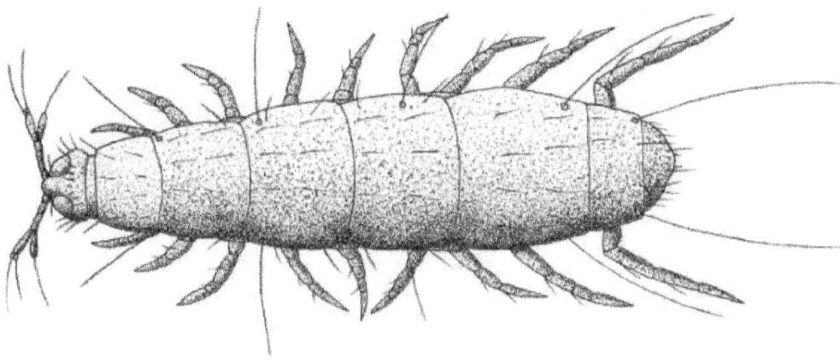

Pauropus huyxleyi, a pauropodan.

Pauropoda is another small group of small myriapods. They are typically 0.5–2.0 mm long and live in the soil on all continents except Antarctica. Over 700 species have been described. They are believed to be the sister group to millipedes, and have the dorsal tergites fused across pairs of segments, similar to the more complete fusion of segments seen in millipedes.

Arthropleuridea

Arthropleurideans were ancient myriapods that are now extinct, known from the late Silurian to the Permian. The most famous members are from the genus *Arthropleura*, which was a giant, probably herbivorous, animal that could be up to 3 metres (10 ft) long, but the group also includes species less than 1 cm (0.39 in). Arthropleuridea was historically considered a distinct class of myriapods, but since 2000 scientific consensus has viewed the group as a subset of millipedes, although the relationship of arthropleurideans to other millipedes and to each other is debated.

Myriapod Relationships

A variety of groupings (clades) of the myriapod classes have been proposed, some of which are mutually exclusive, and all of which represent hypotheses of evolutionary relationships. Traditional relationships supported by morphological similarities (anatomical or developmental similarities) are challenged by newer relationships supported by molecular evidence (including DNA sequence and amino acid similarities).

- **Dignatha** (also called Collifera) is a clade consisting of millipedes and pauropods, and is supported by morphological similarities including the presence of a gnathochilarium (a modified jaw and plate apparatus) and a collum, a legless segment behind the head.

- **Trignatha** (also called Atelopoda) is a grouping of centipedes and symphylans, united by similarities of mouthparts.
- **Progoneata** is a group encompassing Dignatha and symphylans while excluding centipedes. Shared features include reproductive openings (gonopores) behind the second body segment, and sensory hairs (trichobothria) with a bulb-like swelling.
- **Edafopoda** is a grouping of symphylans and pauropodans that is supported by shared genetic sequences, yet conflicts with Dignatha and Trignatha.

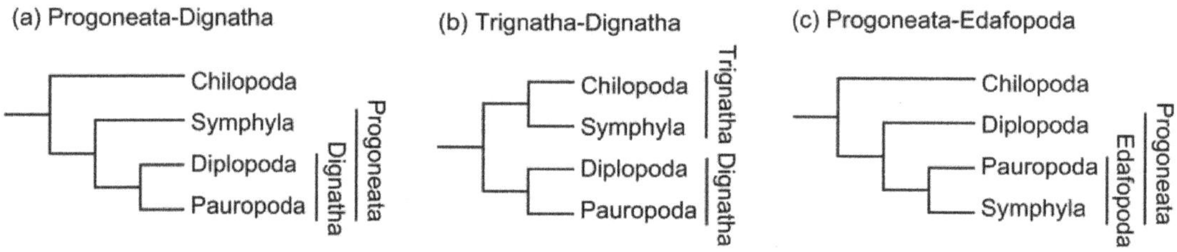

Some of the various hypotheses of myriapod phylogeny. Morphological studies (trees a and b) support a sister grouping of Diplopoda and Pauropoda, while studies of DNA or amino acid similarities suggest a variety of different relationships, including the relationship of Pauropoda and Symphyla in tree c.

REFERENCES

1. Myriapoda". Integrated Taxonomic Information System.
2. Ben Waggoner (February 21, 1996). "Introduction to the Myriapoda". University of California, Berkeley.
3. Paul E. Marek & Jason E. Bond (June 8, 2006). "Biodiversity hotspots: rediscovery of the world's leggiest animal". Nature. **441** *(7094): 707. Bibcode*:2006Natur.441..707M. doi:10.1038/441707a. PMID 16760967.
4. Markus Friedrich & Diethard Tautz (2002). Ribosomal DNA phylogeny of the major extant arthropod classes and the evolution of myriapods. Nature. **376** *(6536): 165–167. Bibcode*:1995Natur.376..165F. doi:10.1038/376165a0. PMID 7603566.
5. Rowland Shelley & Paul Marek (March 1, 2005). Millipede Fossils. East Carolina University.
6. Garwood, Russell J.; Edgecombe, Gregory D. (September 2011). Early Terrestrial Animals, Evolution, and Uncertainty. Evolution: Education and Outreach. New York: Springer Science+Business Media. **4** *(3): 489–501.* doi:10.1007/s12052-011-0357-y. ISSN 1936-6426. Retrieved 2015-07-21.
7. Sue Hubbell (2000). Waiting for Aphrodite: Journeys Into the Time Before Bones. Houghton Mifflin Harcourt. p. 53. ISBN 978-0-618-05684-2.
8. Robert D. Barnes (1982). Invertebrate Zoology. Philadelphia, PA: Holt-Saunders

International. pp. 810–827. ISBN 0-03-056747-5.

9. Myriapod. Britannica Concise Encyclopedia.

10. Barber, A.D. (2009). Littoral myriapods: a review (PDF). Soil Organisms. **81** *(3): 735–760.*

11. Barber, A.D. (Ed) (2013). World Database of Littoral Myriapoda. World Register of Marine Species. Retrieved 25 October 2013.

12. Strange and Unusual Millipedes. herper.com. Retrieved July 2, 2007.

13. Sean P. Bush; Bradley O. King; Robert L. Norris; Scott A. Stockwell (2001). Centipede envenomation. Wilderness & Environmental Medicine. **12** *(2): 93–99.* doi:10.1580/1080-6032(2001)012[0093:CE]2.0.CO;2. PMID 11434497.

14. Gregory D. Edgecombe (2004). Morphological data, extant Myriapoda, and the myriapod stem-group. Contributions to Zoology. **73** *(3): 207–252.*

15. Alexandre Hassanin (2006). Phylogeny of Arthropoda inferred from mitochondrial sequences: strategies for limiting the misleading effects of multiple changes in pattern and rates of substitution. Molecular Phylogenetics and Evolution. **38** *(1): 100–116. doi:10.1016/j.ympev.2005.09.012. PMID 16290034.*

16. A. D. Chapman (2005). Numbers of Living Species in Australia and the World (PDF). Department of the Environment and Heritage. p. 23. ISBN 0-642-56850-2. Archived from the original (PDF) on 25 October 2007. Retrieved 1 January 2011.

17. Jerome C. Regier, Heather M. Wilson & Jeffrey W. Shultz (2005). Phylogenetic analysis of Myriapoda using three nuclear protein-coding genes. Molecular Phylogenetics and Evolution. **34** *(1): 147–158. doi:10.1016/j.ympev.2004.09.005. PMID 15579388.*

18. Central Park survey finds new centipede. American Museum of Natural History. January 29, 2003.

19. Garden Symphylans. Integrated Pest Management on Peppermint-IPMP3.0. Oregon State University. Retrieved July 2, 2007.

20. Pauropods: Pauropoda. Insects and Spiders Scientific Reference. Retrieved July 2, 2007.

21. David Kendall (June 6, 2005). Pauropods & Symphylids. Kendall Bioresearch.

22. Wilson, Heather M.; Shear, William A. (2000). Microdecemplicida, a new order of minute arthropleurideans (Arthropoda: Myriapoda) from the Devonian of New York State, U.S.A.. Transactions of the Royal Society of Edinburgh: Earth Sciences. **90** *(4): 351–375.* doi:10.1017/S0263593300002674.

23. Shear, William A.; Edgecombe, Gregory D. (2010). The geological record and phylogeny of the Myriapoda. Arthropod Structure & Development. **39** *(2-3): 174–190. doi:10.1016/j.asd.2009.11.002. PMID 19944188.*

24. Edgecombe GD, Giribet G. (2002). Myriapod phylogeny and the relationships of Chilopoda. In J Llorente Bousquets, JJ Morrone, HP Ulloa. Biodiversidad, Taxonomia y Biogeografia de Artropodos de Mexico: Hacia una S´ıntesis de su Conocimiento *(PDF). Univ.* Nac. Aut´on .M´exico: Prensas Ciencias. pp. 143–168.

25. Miyazawa, Hideyuki; Ueda, Chiaki; Yahata, Kensuke; Su, Zhi-Hui (2014). Molecular phylogeny of Myriapoda provides insights into evolutionary patterns of the mode in post-embryonic development. Scientific Reports. **4** *(4127).* doi:10.1038/srep04127.

26. Zwick, Andreas; Regier, Jerome C.; Zwickl, Derrick J.; Gadagkar, Sudhindra R. (2012). Resolving Discrepancy between Nucleotides and Amino Acids in Deep-Level Arthropod Phylogenomics: Differentiating Serine Codons in 21-Amino-Acid Models. PLoS ONE. **7** *(11): e47450. doi:10.1371/journal.pone.0047450.*

CHAPTER 8

Lepidopterology

Chapter Contents

- **Introoduction**
- **History**
- **Lepidopterological Societies**
- **Melittology**

Llepidopterology the scientific study of butterflies and moths (the lepidoptera) while butterfly is a flying insect of the order lepidoptera, distinguished from moths by their diurnal activity and generally brighter colouring.

INTROODUCTION

Lepidopterology is a branch of entomology concerning the scientific study of moths and the three superfamilies of butterflies. Someone that studies in this field is a lepidopterist (or aurelian).

Danaus plexippus and *Actias luna*.

HISTORY

Post-Renaissance, the rise of the "lepidopterist" can be attributed to the expanding interest in science, nature and the surroundings. When Linnaeus wrote the tenth edition of the *Systema Naturae* in 1758, there was already "a substantial body of published work on Lepidopteran natural history" (Kristensen, 1999).

These included:

- *Insectorum sive Minimorum Animalium Theatrum* – Thomas Mouffet (1634)
- *Metamorphosis Naturalis* – Jan Goedart (1662–67)
- *Metamorphosis insectorum Surinamensium* – Maria S. Merian (1705), whose work included illustrated accounts of European Lepidoptera
- *Historia Insectorum* – John Ray (1710)
- *Papilionum Brittaniae icones* – James Petiver (1717)

LEPIDOPTEROLOGICAL SOCIETIES

Lepidopterists are served by a number of scientific societies, both national and international in scope. They promote research in lepidopterology and dissemination of the findings therefrom primarily through the arrangement of talks such as the biennial European Congresses of Lepidopterology or the TILS Leps Talk. These societies include:

- Lepidopterists' Society
- Societas Europaea Lepidopterologica
- Lepidoptera Research Foundation
- North American Butterfly Association
- Association for Tropical Lepidoptera
- International Lepidoptera Survey
- Lepidopterological Society of Denmark
- Lepidopterological Society of Finland
- Lepidopterological Society of Japan
- Sicilian Lepidopterological Association
- Southern Lepidopterists' Society
- Study Group of Hessian Lepidopterologists
- Lepidopterists' Society of Africa

MELITTOLOGY

A bee drinking water.

Melittology (from Greek μέλιττα, *melitta*, "bee"; and -λογία *-logia*) is a branch of entomology concerning the scientific study of bees. Melittology covers the species found in the clade Anthophila within the superfamily Apoidea, comprising more than 20,000 species, including bumblebees and honey bees.

Subdivisions

- **Apiology** - (from Latin *apis*, "bee"; and Ancient Greek -λογία, *-logia*) is the scientific study of honey bees. Honey bees are often chosen as a study group to answer questions on the evolution of social systems.
- **Apidology** is a variant spelling of *apiology* used outside of the Western Hemisphere, primarily in Europe; it is sometimes used interchangeably with *melittology*.

- **Apicology** - the study of honey bee ecology.

Melittological Societies

Melittologists and apiologists are served by a number of scientific societies, both national and international in scope. Their main role is to encourage the study of bees and apicultural research

- *International Bee Research Association*
- *National Bee Association of New Zealand*
- *British Beekeepers Association*
- *German Beekeepers Association*
- Stephen Taber III, (1924-2008), Innovator in the practice of artificial insemination of queen bees for the purpose of developing disease resistant and gentle bee colonies.
- Mark Winston, Studies life history, caste structure, and reproduction in social insects and pheromones of honey bees at Simon Fraser University.

List of notable Melittologists and Apiologists

- Freiderich August Bechly (Fred Bechly), (1835-1916), was a correspondent for the American Bee Journal.
- Charles Butler, (1560-1647), early English beekeeper and researcher.
- Charles Dadant, (1817-1902), Modernized beekeeping.
- Jan Dzierzon, (1811-1906), Discovered parthenogenesis among bees, proposed first sex determining mechanism for any species.
- Savannah Foley, studies genetics and communication over long distances at the University of South Florida, leading a team investigating recent dropping numbers of honey bees.
- Michael S. Engel, (b. 1971), studies honey bee and other bee taxonomy and paleontology at the University of Kansas.
- Karl von Frisch, (1886-1982), Nobel Prize winner, studied honey bee communication.
- Robert A. Holekamp, (1848-1922), Early urban apiculturalist and advocate.
- Jay Hosler, Professor at Juniata College, Author of the award-winning comic *Clan Apis*.
- François Huber (1750-1831), Swiss naturalists, introduced the "Ruche en livre» (Book hive), a beehive system in the shape of a book, first published in «Nouvelles observations sur les abeilles, adressées à Charles Bonnet», 1792 (New observations on the natural history of bees).
- Karl Kehrle (a.k.a. "Brother Adam") (1898-1996), Benedictine monk, beekeeper, and an authority on bee breeding, developer of the Buckfast bee.

- Warwick Estevam Kerr, (b. 1922), Studies genetics and sex determination in honey bees. Responsible for introduction of Africanized bees to America.
- William Kirby, (1759-1850), Author of the first scientific treatise on English bees.
- L. L. Langstroth, (1810-1895), Modernized American beekeeping.
- Martin Lindauer, (1918-2008), studied communication systems in various species of social bees including stingless bees and honey bees.
- Sir John Lubbock (the 1st Lord and Baron Avebury) (1834–1913), wrote on hymenoptera sense organs.
- Robert E. Page, Jr., Studies population genetics and the evolution of complex social behavior at Arizona State University.
- Petro Prokopovych, (1775–1850), Ukrainian beekeeper, founder of commercial beekeeping.
- Moses Quinby, (1810-1875), Early American commercial beekeeper. Invented modern bee smoker.
- Gene E. Robinson, Studies mechanisms of bee-havior at the University of Illinois.
- Amos Ives Root (1839 – 1923), Innovator in honey harvesting techniques. Published first account of Wright brothers flight in his beekeeping journal.
- Grace Sandhouse
- Justin O. Schmidt, Studies bee nutrition, chemical communication, physiology, ecology and behavior. Created Schmidt Sting Pain Index.
- Thomas D. Seeley, Studies group organization using the honey bee as a model system at Cornell University.
- Robert Evans Snodgrass, (1875-1962), Author of one of the first comprehensive books on honey bee anatomy and physiology.
- Stephen Taber III, (1924-2008), Innovator in the practice of artificial insemination of queen bees for the purpose of developing disease resistant and gentle bee colonies.
- Mark Winston, Studies life history, caste structure, and reproduction in social insects and pheromones of honey bees at Simon Fraser University.

REFERENCES

1. Harper, Douglas. lepidoptera. The Online Etymology Dictionary. Retrieved 8 February 2011.
2. Kristensen, Niels P. (1999). Historical Introduction. In Kristensen, Niels P. Lepidoptera, moths and butterflies: Evolution, Systematics and Biogeography. Volume 4, Part 35 of Handbuch der Zoologie:Eine Naturgeschichte der Stämme des Tierreiches. Arthropoda: Insecta. Walter de Gruyter. p. 1. ISBN 978-3-11-015704-8. Retrieved 30 November 2010.
3. Bees - Facts About Bees - Types of Bees - PestWorldforKids.org». pestworldforkids.org. Retrieved 2016-04-26.

4. http://www.gutenberg.org/files/26457/26457-h/26457-h.htm
5. ASU SoLS Faculty: Robert E. Page
6. University of Illinois - Honey Bee Research Group
7. "Grace Sandhouse Papers". Record Unit 7456. Smithsonian Institution Archives. Retrieved 6 March 2012.

Index